STEPHANIE BRANDOLINI

BRIDGING THE DIGITAL DIVIDE

ISBN: 978-1-960136-61-9

Introduction

Stephanie Brandolini is a best-selling author, award-winning screenwriter, filmmaker, speaker and high ticket business coach from Vancouver, British Columbia, Canada. After graduating from film school, Stephanie worked her way up in the visual effects production world, gaining in-depth knowledge of the TV and film industry. Through her success and spirituality, Stephanie discovered a deeper calling to go all in on her writing and creative gifts and serve others.

Now, as a creative entrepreneur, she is on a mission to help driven individuals and families break free from the matrix, uplevel their health and finances, get their time back, and build a legacy. All while creating and collaborating on writing, film, and speaking projects that uplift, inspire and transform audiences worldwide.

Bridging the Digital Divide was inspired by the 2023 writers' and actors' strikes and the rise of AI technology. This novel blends real emotions with Stephanie's deep knowledge of psychology, metaphysics, transcendence, spiritual work, and energy healing. It explores the intricate intersections of technology, spirituality, and human consciousness, delving into the complexities of sexuality, addiction, and self-discovery.

In this book, you will follow Samantha Sands as she embarks on this transformative journey.

Through her story, you'll find inspiration to ground yourself in your passions and gifts, and discover how to transmute what you might perceive as your greatest weakness into your greatest strength.

Join Stephanie's community of dreamers and achievers, and embark on your own journey of self-discovery and empowerment.

With love, peace, and blessings,
www.stephaniebrandolini.com

Table of Contents

CHAPTER 1

T he unsettling orb on my desk glowed rhythmically, projecting a holographic screen before me.

I watched as letters formed words, mounting into lines of text. Each one a reminder that I would have written this way better myself.

Seriously? This is why my writing room was laid off?

The orb's light show continued as if pulsing the words out. This was IRIS, the Writers' Room Bot that was forced upon me weeks ago, and fucked my world in the ass. Its pulsing glow slowed and faded to the thing's natural (rose gold of all colors) state as it finished its last sentence like a ghost had punctuated the page.

- *Outline of Digital Dystopia complete,* *-* IRIS declared. In her smug, cunty tone.

"Upload to Tablet Presenter Two," I commanded.

The second tablet in front of me lit up immediately. Despite all this AI bullshit, the bot really did have uncanny data transfer rates. Maybe that part could be salvaged from this mess of a deal the Board had gotten us into.

And speaking of...

I checked myself in my floor-length mirror.

Pressed power suit? Check. Don't-fuck-with-me hair? Check. Overall expression...too dour, but we can fix that.

I imagined gears in my mouth grinding to turn the corners...up.

Nailed it.

I snatched the tablets and strode to the door with the poise and purpose of Lady Gaga claiming the dance floor she fought for.

I was almost out the door when I heard...

- You're welcome, Miss Sands. *-*

I glanced back. The orb sat there, its holographic screen suspended in the air like a glowing white canvas.

For no reason I could then fathom, I got the sense of an abandoned child yearning for praise. Something in my heart hitched—I choked up. Tears welled—

SMACK! I slapped my own goddamn face.

It felt good.

The orb's holographic screen blinked out of existence, leaving only darkness as I slammed the door.

***

I stomped down the ostentatious hallway of Sands Studios' executive-level offices.

Skylights galore beamed sunlight on me. Light that would have served the creativity of my writers and I, but oh no, the damn Board had to impress clients—those that deigned to get off their remotely-working asses anyway.

What a fucking farce.

CLICK CLACK CLICK—I loved the sound of my Louboutins on marble—Boss Bitch Mode activated!

Posters of so many TV shows I'd run blurred past me like hungry ghosts, reminding me of a past that would never be again. A past where I was the creator, not some know-it-all machine.

A gold plaque on the double doors ahead glinted my uncle's name back at me:

JEFFERSON SANDS

- CEO -

A name that evoked a coddling protection I wanted to wrap myself up in like a security blanket and hack away at with a machete to free myself from—all at the same time.

Ain't dichotomy grand?

Cindy, his Instagram-pretty assistant, braced herself at my approach. Cue the fake plastic smile...there it is. I pretended to be immersed in my phone as I steered toward the boardroom.

This chick just doesn't get boundaries though. "Hi Samantha, the Board is ready for you."

"The building's got my name on it, Cindy. The Board's ready

when I get there. But thanks! Oh, and it's Miss Sands, please."

I slung my own fake plastic smile back at her. Clash of the passive-aggressive bullshit.

"Of course, Miss Sands. Sorry about that."

She actually looked wounded beneath her pristine contouring.

Maybe I was too harsh...

Then I heard the rumblings of muted conversations behind the grand double doors and clenched my fist. No time to get soft when you're heading into battle. I steeled myself and recited my inner mantra, my contract with myself:

I am a confident, creative, powerful leader—I am a confident, creative, powerful leader—I am a confident, creative, powerful leader...

And owned it as I strode through the doors.

CHAPTER 2

I stood at the head of the boardroom table that seemed to stretch on like the path to an executioner's block.

At the end of my green mile, Chairman Ross loomed like a judgemental vulture in an overpriced suit. He was flanked on either side by the Board who collectively had more power over the production company my father built than I did. I refused to let their scrutiny derail me as I continued my Oscar-worthy presentation.

Uncle Jeff sat on my right, beaming up at me. Despite the cinched-up suit he had on today, he would always be my Yoda, training me in the ways of business politico bullshit.

I caught his eye with our shared secret smirk when a cold, nasty energy prickled my senses from the opposite end of the table. Chairman Ross' steely gray eyes bore into mine under a furrowed brow, etching deep frown lines that not even the best Botox could fix. Thank the Universe I had Uncle Jeff's warmth at my side, combating the icy presence of this asshole that felt like a slap in the face.

But I wasn't here to back down.

Ignoring him, I gestured to the presentation of the outline I'd done myself versus the one She-Who-Shall-Not-Be-Named generated from my prompts.

"As you can see, the test I did to write an outline of our next show using AI technology clearly shows the inadequacies of such an endeavor. Not to mention the damage to creativity itself. Why are we watching shows if not to connect with the human spirit who created them?"

My pause for dramatic effect landed with about seventy percent of the room. Of course, the pro-bot bastards had narrowed gazes and insulted expressions. All they gave a shit about was, you guessed it, motherfucking money.

Chairman Ross scoffed. "Creativity is subjective, Miss Sands. Despite your highly-coveted 'process...'" The dickhead actually had the nerve to air-quote that. I gritted my teeth and bore it behind a tense smile. "... our competitors have outperformed us in both ratings and revenue for the third quarter in a row."

"You mean sellouts like BotFlix? Have you actually watched the trash they're churning out? It's formulaic drivel that's frankly an insult to viewers and creatives alike. My father built this company with a reputation for quality, and I will *not* let it become just another content mill."

"Quality is also subjective; numbers are not. Our ROI has plummeted, our engagement metrics are down, and even our international sales have taken a hit. If we don't adapt, we risk not just relevance but the studio's solvency."

I froze as the abject fear of being part of what might drive our studio into the ground hit me like a self-inflicted gut punch. I recovered quickly, summoning my energy to flow through the fear as Uncle Jeff had taught me.

Not quick enough to escape Ross' flinty eyes though. He smiled like a wolf who'd cornered prey. "The multi-million dollar AI integration partnership I sourced with Falls Tech is the answer to all our problems in this matter. Unless you have a better idea of course?"

The Board members looked at me. Those who'd been a part of the company since I was a child had hope and trust in their eyes. Unfortunately, they were the minority. The majority were all Ross' cronies, "sourced" by the man himself.

I was close to cracking when that vomit-worthy machine's creator, Brad Falls, flashed me a smile that lit up the room.

He was annoyingly good-looking. The kind of man who no doubt used it to fuck women over in more ways than one.

I felt my face trying to express my disdain but reined it into calm, casual composure. No way was this fuckboy messing me up when I was so close to glory.

"I propose we strike the AI from our room and bring back my— excuse me—*our* family of talented writers who will make our next show, *Digital Dystopia*, the most heartfelt, engaging, and lucrative yet. It'll be a revolutionary series that sticks to the values of true creativity in a world that seems to be forgetting what the truth really is."

There was a palpable pause. The board members exchanged glances, some nodding in agreement, while others remained skeptical. They whispered among themselves, sounding like a swarm of readying locusts.

I could feel the weight of so many decisions hanging in the air. I held strong to my contract as my heart began a deflating descent into that dark spiral that always beckoned me...

Not today, Satan! I summoned my energy, imagining it radiating out for the best possible outcome when suddenly—

CLAP CLAP CLAP—Brad aroused the room's attention as he stood up at the opposite end of the table.

I quirked a brow. If this dude was trying to face off against me, it was game on.

"Thank you for your impassioned stance, Miss Sands. We all respect your dedication to this studio and your craft of storytelling. No one doubts that a writer of your caliber can outperform IRIS." He paused for effect, all ease and charm while I tensed up, ready to rumble. "But let me ask, did you try the 'Getting To Know You' tool before testing IRIS?"

I'm not getting to know your fucking bot, you creativity-busting douche!

Out loud I said, "I did not."

"That's likely where this test went wrong. Not a problem though. I'd be happy to walk you through the process." He gestured to the Board with a sunny smile that reeked of bullshit. "Board members,

I propose we temporarily suspend Miss Sands' motion. Instead, we do a trial. Let's see what IRIS can truly achieve once she's better acquainted with Miss Sands." He looked me right in the eye as he concluded, "I assure you, it'll be worth the exercise."

You and your bot don't know the first thing about the time it takes to create anything that matters!

A placid smile was all I let show as murmurs rippled around the room between the board members. I felt the divide before I saw it—half kept their hands lowered in favor of my motion, and half had raised their hands for Brad's.

Those in agreement with me received silent nods of gratitude.

I kept my reactions in check for the others.

Tension mounted as we all turned to the deciding voter, Uncle Jeff. I caught his eye, using mine to try and sway him away from this Faustian deal.

He gave me that smile he used to give me as a teen when I'd get extra detention for arguing with my teachers over the detention I'd already gotten for skipping school.

Even after all this time, only the bullies and I knew the real reason behind my many after-school sessions...

Suddenly, I blinked and it was like the world sped up around me like a weird reverse of

lightspeed from the original Star Wars trilogy my dad and I would nerd out on.

I shook my head and found myself still staring into Uncle Jeff's bright green eyes, so like my own. Something felt strange, like I'd been lost for a while in my high school drama déjà vu.

That reverie was immediately obliterated when Uncle Jeff raised his hand.

My heart sank—

Straight into that despairing spiral.

CHAPTER 3

"What the fuck, Uncle Jeff?"

We were in the sanctity of his office now, an intimate space I once loved to play hide-and-seek in. I could almost see my younger self peeking around the sunlit curtain—worst hiding spot ever, by the way. Behind the curtain, a grand bay window formed a deep recess, its mysteries hinted at by the sun's gentle glow.

Uncle Jeff emerged from his personal bathroom wearing a Japanese-style kimono—all five feet of him.

"Thank the Universe you're back to your Danny Devito of business shamans' look. Suits stifle your shine."

"Sometimes we need to play their game so we can play ours."

"How about when we're losing?"

"The tides have brought us here, Samsey. We're Sandses, and what do we do?"

I locked eyes with him. This was our sacred Sands' Mantra we'd crafted together long ago. When the tides brought us so much devastation and we were forced to...

"Shift with the ever-changing flow," I declared. Then I bit the inside of my cheek, holding back a flood of dark memories.

He took my hands and closed his eyes. I followed suit. I could feel the energy running through us. Our sacred connection—our human connection. I gripped his hands tighter as the flow thrummed, peaked, and...

Released.

I let out a huge breath I didn't even know I was holding in. Then opened my eyes to find Uncle Jeff's pure green gaze staring right into me. His eyes had the same hue as mine...as my dad's had been...

I felt a serene sense of upliftment. It began to buoy me up when suddenly, Ross' snide eyes flashed in my mind and my buoyancy morphed back into a cluster of consternation I couldn't seem to shake.

"But seriously!" I declared, throwing my hands up in the air. "Chairman Ross is compromising our vision. This—" I gestured to our exchange, "—kind of energy is what built this company and it doesn't exist with that insidious device. Fuck the financial shitstorm we're in! Sell my shares of the company if that's what it takes!"

Uncle Jeff winced. "It's not that simple, Samsey. Those shares are what give you a legal voice in that room."

"What good is my voice when the high-and-mighty Chairman is against me? Why do we even have a Chairman like him anyway? He's been a piece of shit to me since my first show here."

Uncle Jeff chose his next words carefully. "Charles wasn't always so...unpleasant." He paused, a small smile played at his lips before firming up into a frown. "People change, Samsey. Priorities shift."

I caught a flash of something in his eyes. "What aren't you telling me?"

He waved a hand. "Nothing. I'll speak to Charles about his attitude toward you, but know that he's doing his job to ensure our company thrives in the long run."

"But what's the point of thriving if we lose our souls in the process? We need to remember why we started this company—for the stories, for the magic!"

"Movie magic doesn't exist without money. If you truly want to stand for our vision, you'll need to learn the financials and negotiations that I deal with as CEO."

"Fuck no! I'm a storyteller, not a number-crunching sales machine. Just don't take away the essence that me, you, and Dad—"

I stopped short, choked up on too much emotion.

I hid my clenched fist in the folds of my jumpsuit, my French manicure digging like daggers into my palms.

The pain staunched the tears.

"You're out of alignment with your Merkaba. A session with Shiba would have prevented that."

"Ugh, if you push her on me one more time, Uncle Jeff, I swear—"

"Your reactivity only further proves my point. Watch, I can prove her to you..."

He was already going for his phone as a jolt of angst sizzled through me about how much time that would take. Time I didn't have if I was going to save my career, my writers' careers, my family's company...

I felt the familiar spiral of overwhelm threaten to pull me under.

On the outside I smiled brightly, drawing from my inner child's innocence I knew Uncle Jeff couldn't say no to. "You're my Merkaba Master, we don't need Shiba. You're right though. No wonder I've been feeling so out of it. Can you teach me how to get back into alignment again, please?"

His eyes brightened like moonlight shot through aquamarine. Then he sighed and abandoned his mission. "Only if you promise you'll have a session with her one day. I'm far from master level on my own spiritual healing journey. Shiba can see misalignments I'm sure to miss and she can do things for you I can't."

I rolled my eyes on the inside. "I don't want some stranger teaching me these techniques."

"She won't be a stranger once you meet her."

I couldn't help but laugh. "Touché, Uncle Jeff. Touché."

He just looked at me.

I scoffed and tapped my throbbing temple. Didn't cut it so I started walking back and forth. Soon my headache ebbed enough for me to say, "Fine! I promise I'll get in touch with her after I "get to know" that stupid fucking machine, prove its incompetence, get our writers hired back..." I was pacing a hole in the carpet at this point, listing off my laundry list on each finger.

"Samantha..."

I barely heard him. All the mounting to-dos spewing from me like an angry avalanche. "...write the comeback show of my career, unspool my new show ideas for next year's quarter, then the next and the next and—"

"Samantha!" He did something with his voice he'd done before, but this time it was stronger. It was like the decibels were wrapped in a resonance that struck me at my core. I ceased my pacing, my listing, my overthinking.

"Merkaba realignment lesson number one: slow down."

I walked over to him as if floating, meeting him in the center of his office as he continued speaking, "And remember, it's not just a tool to calm you down. Actually, you'll love this. I recently learned from Shiba that Merkaba means 'chariot' in Hebrew. So you can think of it like an energetic chariot to your higher self. Even deeper in the language aspect, the word Merkaba also comes from the Old Testament in the good old book where this prophet named Ezekiel had a vision of the throne of God."

I paused, starting to imagine that, then scoffed and quirked a brow instead. Too irked to let the flow in. "How can we be sure he wasn't on something?"

Uncle Jeff just looked at me, mirroring my raised brow.

"I'm just kidding, calm down."

Now it was Uncle Jeff's turn to scoff. "Samantha, I am calm. How are you?" He nodded to my fidgeting fingers and restless right leg, bouncing up and down as if vibrating.

I scowled and forced my body rigid. "Ugh, OK, point taken. I get that we're all energy and this Merkaba thing has been helping me be more Jedi over Sith, but seriously, Uncle Jeff, New or Old Testament, that 'good old book' has never yielded me anything but confusion and disappointment."

His eyes softened, no doubt lost in long-ago memories. "The world has evolved exponentially since it was written, Samsey. Of course not all of it is fact but I take from it what serves and leave the rest." He nodded to the other side of the room. "This calls for a Zen Den moment. Come along." He said, leading me in his debonair way to the bay window that was veiled from view.

And through the curtain...

A spectacular circular view of the Big Apple herself. Two meditation chairs and hanging sacred geometry tapestries adorned the space.

Ah, the Zen Dome, a place to let the glory of our city wash away the bullshit of it.

I sat on my chair. Ready to go.

He looked at me, eyes slightly unfocused as he scanned my aura.

"We've both picked up negative energy from that board meeting. How about something to take the edge off?"

He ducked back inside. I already knew where he was going: his treasure table, as I'd coined it. Basically a stocked-to-shit liquor cabinet on which gorgeous enameled boxes lay, ready and waiting. One was full of cigarettes, one full of weed, and, to make it a trifecta, one full of Colombia's finest stardust.

I smirked, anticipating him coming back in a huff when he realized I'd flushed his stash down the toilet.

When he waltzed back in with the weed box I fumed. "Hey! I drowned those treasures! Doc's orders! You're not giving yourself another heart attack as long as I'm around. Give that here."

"Orders don't apply when you're connected to the divine. It's my time when The Creator wills it."

"Oh yeah, because 'The Creator' has been so fair to us in that regard?"

He at least had the decency to look mildly abashed. I saw my in and I took it.

"Is that Shiba-approved?" I asked, sneering at the joint and jonesing for one myself.

"Shiba's word isn't law. The point of anything you learn is to make it your own. Besides, high-functioning types like us need a little extra to slow down. There's a reason why we jones for it, Samsey, because we need it."

He opened the beautifully enameled box, rows of perfectly rolled joints met my eye.

I picked one up. "This is all you've got in your treasure table though, right?"

His eyes flicked off to the side. It was slight as fuck, but I caught it.

"I fucking knew it! No matter what divine Jedi mind tricks you learn you'll still never poker face me. Where's the stardust? I won't be a culprit in your debauchery again."

"Oh come on, Samsey. I know my limit now and I stay within it."

"Fuck you. I cut back in solidarity."

"I cut back too!"

I just stared him down.

"What do you want from me, Sam? I'm not going for sainthood here."

I laughed at that, imagining him in priestly regalia. "I sure as shit hope not."

We had a little mock standoff, but I couldn't suppress my smile. Uncle Jeff had always put the cool in cool uncle. "So you got a light or what?"

He did.

We sat in silence and smoked, hotboxing the Zen Dome. Watching the craze of New York City fly by through the haze of smoke was actually the most serene I'd felt in months.

Just then, the angst of wasting time came to greet my peaceful thoughts.

You have way too much to do, you nepotistic idiot. What do you think you're doing with this garbage?

"OK Samsey, begin the breathing." Uncle Jeff's comforting voice broke through the nasty noise in my head.

Fucking finally.

I breathed in long, deep, and as slow as I was capable. With each inhale and exhale, I felt a shift, as if a heavy weight was lifting.

Along with it, I felt a begrudging sense of allowance. A big part of me didn't want to relax into this. But I kept breathing and after what felt like an eternity, the darkness in my mind's eye unfurled, revealing an ethereal outline of my body floating in the abyss of my imagination.

"Visualize your upper tetrahedron, like a pyramid of glowing golden light," he murmured.

I did, watching on the inside as it shimmered to life around me. It spun gracefully, cocooning my upper body in a familiar embrace. That part of me that didn't want to relax finally did as I felt an overwhelming flood of gratitude for it rush through me.

"Now, bring forth the lower one," he continued.

A second pyramid, this one inverted and green, wrapped around my lower half. Unlike its counterpart, it flickered uncertainly, resembling a faulty neon sign. I suddenly felt restless, but held on to the gratitude as it slowly started to spin. There was nothing graceful about this one as it fizzled and sparked, feeling like a rusty gear I had to crank to move.

I didn't know what to do so I just let it be. There was no stopping this kind of process once it started.

Both tetrahedrons spun, faster and faster. My upper golden pyramid flowed smoothly but my lower one jolted and jittered, frazzling fractals spinning out below me.

As they spun, beams of light pierced through me, converging at my heart. A rush of warmth met the striking ice of resistant fear, like something was barring these energies from melding seamlessly. I gasped as some of the light struck me, as if through a sieve that was sifting out most of what I truly needed.

The realization of what I was missing out on brought tears to my eyes as I slowed down the process, letting my Merkaba dissolve from my mind's eye.

I blinked my eyes open, pushing the tears away. Through their haze, Uncle Jeff's gaze seemed to peer straight into me. "Are you OK, Samsey?"

Wasn't that a loaded question?

"I am," I replied, my voice thick. I cleared my throat and postured up, so he wouldn't see just how affected I really was. "Thank you. I needed that for sure. Merkaba's back in alignment, Master Jeff. The Force is again strong with me." I backed up my half-truth with a weak smile.

He laughed as he pulled his chair over and sat down next to me. I leaned into his comforting embrace and we smoked until the sun went down.

It was a glorious respite I wouldn't have shared with anyone else.

CHAPTER 4

A nd just like that, my zen was gone.

I glared up at the holographic screen that was prompting me to answer another stupid fucking question. Brad sat next to me, chirping away with his AI nonsense.

"Think of her like one of the countless *Times* interviewers you've sat down with. It's the same thing."

I raised my eyebrows sky-high.

"OK, not the same thing, but the same idea. Remember, IRIS is an Intuitive Response Ideation System. Going through this mode will—"

"The kind of intuitive responses I need in a writers' room can't come from a machine, Mr. Falls."

"How do you know that?"

"That's my intuitive response, it defies explanation."

"Sounds more like it's defying possibility."

The raging bitch in me reacted hard.

Who the fuck does this douchebag think he is? Seriously, Sam, let's trash the damn bot, book it to the bar, and write whatever the fuck we want there!

I took a deep breath to rein her in before I continued.

"Why is this necessary? I fed it my writing samples."

More like you sold your soul to the devil himself.

Yeah, she's got a flair for the dramatic.

"As a showrunner, have you ever hired anyone based solely on their writing samples?"

"Of course not."

"Why?"

I paused, chewing my lower lip as I formed my answer. "I need to meet them and get a sense of who they are to really see if they're a good fit for the show and the room."

"That's just like IRIS' algorithm in this mode. You get to give her that sense of who you are and what you want so she can support you."

"As opposed to a room of human beings creating what we're psychologically wired for?"

"What do you think IRIS is wired for?"

I rolled my eyes and sighed long and hard. Unprofessional, but I couldn't help it.

"Look, Samantha—"

"Let's stick with Miss Sands."

"Fine. Miss Sands, what if you let go of thinking of IRIS as a threat and started thinking of her as an ally?"

"I don't need—"

"She's not going to become the driving force behind your projects, I assure you. My intent is and always will be for that to come from the showrunner and writers. Years of psychological and philosophical study has gone into building her into an artificial intelligence that supports creativity. She actually can't generate a story on her own. Her output depends upon the user's input. You're the best there is in my eyes, which is why I thought you'd be the best to test her, but if you're this resistant maybe we should take our funding elsewhere."

He rolled his chair away and strode to the door.

I glared at the makes-me-want-to-drink machine.

He seriously expects you to create with that thing? Blasphemous! Your dad's gotta be rolling hard in his grave right now—and not the ecstatic kind.

My eyes flicked to the face-down picture on my desk. The one I hadn't been able to look at while this clusterfuck was going on.

I picked it up. My fingers grazed across the glass that encased a picture of Seven-Year-Old Me, Uncle Jeff, and...

My dad, Jasper Sands.

He towered over Uncle Jeff and I like the gentle giant he was— gentle until anyone messed with us, that is. He wore his classic Star Wars shirt loud and proud, arm around me wearing my own identical shirt of the iconic franchise's legendary lettering. I knew from memory the infamous scroll decorated the back. Per usual, I had a notebook in hand, scribbling away at something that most definitely couldn't wait for a photo op.

I stared into my dad's eyes. I could almost hear his voice...

"Nothing stays the same forever, Samsey. It's how you shift with the sands of life that defines you."

I put the photo back, face up, and took a deep breath.

"I am the best to test IRIS, Mr. Falls. Let's stick to our contract."

He was halfway out the door when he paused and grinned over his shoulder. I was annoyed at myself for finding it hot.

"Sounds good to me, Miss Sands. And please, call me Brad."

I caught his eye with a begrudging nod as he walked out, still grinning away.

"Ugh, infuriating man!"

Is what I said as the corners of my mouth involuntarily rose and I clenched my thighs together to staunch my sudden wetness.

"Fuck no, Sam. Not with that one."

- *Excuse me, Miss Sands, may I be of assistance?* *-*

I guffawed. Then I actually thought about it.

"Yes, you can DM Cindy to get me a caramel macchiato, with sugar-free syrup, coconut milk, and extra caramel drizzle."

A rainbow-like icon lit up on the screen like the dreaded three dots that hover when a lover may or may not respond. But there was no hovering, no waiting at all, actually. The color spectrum lit up, glowed, and IRIS chimed like a clarion bell.

- *Your order is on its way.* *-*

I raised a brow and pursed my lips. It took me a minute before I could say, "Thank you."

Against my former resistance, my curiosity was spiked like the punch at an underage dance.

CHAPTER 5

Two caramel macchiatos and a gluten-free chocolate croissant later, my cup of curiosity was overflowing as I introduced IRIS to the wonderful world of me.

"OK, let's get one thing straight—I'm single by choice."

- Of course, Miss Sands. I am merely noting the contrast of how the main characters in all your shows steer toward partnership and this ideal that humans have of true love, when being alone is your personal choice.*-*

My bitchy inner self was on high alert.

Does this cunt think she's our shrink now?

I eyed my liquor cabinet, Hendrick's just within reach...

Liquid love will make this whole thing a lot more fun!

I saw my hand move to grab it and clenched my fist instead, relishing the pain of my nails digging into my palms.

To IRIS, I said, "Life is a dichotomous dance. My work explores that in fantastical ways and that bullshit ideal of true love sells."

- Indeed it does. What is the driving force behind the stories you create? *-*

Fuck it!

I grabbed my gin and poured three fingers—nah, let's go four. I took a long sip that was like liquid lubrication for my blocked-up throat chakra.

Alright, IRIS, you asked for it.

"I believe everything humans do stems from love. A need for it, a lack of it...it's fascinating."

That rainbow icon lit up again.

- Fascinating can be described as extremely interesting and capable of arousing and holding attention.*-*

I rolled my eyes and refilled my poison.

"Well, that's an AI-generated response if I ever heard one."

- Of course, Miss Sands. I am highly optimized but still remain an artificial intelligence. I bring up the known definition of fascinating to ask you your opinion on it. *-*

"Yeah, yeah. Well since we're "getting to know each other," let me tell you the most fascinating secret most people won't: Humans will spend the rest of their existence trying to understand love and never be able to." I was getting slurry and I liked it.

- Why is that, Miss Sands? *-*

I downed my glass and through the empty bottom saw the photo of my dad, Uncle Jeff, and Seven-Year-Old Me. I could see only my dad clearly.

I felt like a drunken pirate sighting home through a spyglass. I tried to laugh at my own little joke but choked on tears instead. I clenched my fist but couldn't stop the flood.

Well, fuck me, a machine made me cry. At least she can't see me. I reined myself in enough to speak and faced the photo—staring straight into my dad's eyes again. Eyes so like mine.

"Love can't be understood...only felt."

Saying those words, to a fucking bot no less, broke something open within me. A wall of resistance I didn't even know I'd built was torn asunder by the flood of emotions I felt in that moment.

It was the best and worst feeling of my life.

Up until then at least.

I sat there in my fancy-ass, boss-bitch chair, clutching my precious photo while rocking back and forth. More gin at the ready.

- *Miss Sands, your speech is not computing. How may I support you?* *-*

And just as quickly as the flood had pulled me into that suffocating spiral, levity lifted me higher.

I laughed. Loud and boisterous. Like I had as a kid before I entered the wild world of womanhood and suppressed anything that would make me look ugly.

How sad that all seemed now.

"Just be with me, IRIS...please."

- Of course, *Miss Sands. I am with you.* *-*

I felt gratitude in that moment that superseded all I'd felt before. I couldn't explain it, but I could feel it. Then that feeling, combined with the vortex of all the emotions I'd released, pulled me...

Back—

    Back—

        Back...

To the moments the emotions were rooted in.

I heard myself speaking to IRIS, felt my fingers pick up their inspired dance across her keyboard interface as if I were detached—and I was. I was no longer in my body, I was somewhere else...

Somewhen else.

I'm not actually sure if my eyes were open or closed, but in my mind's eye, I saw the gold and green tetrahedrons of my Merkaba spinning like propellers, supporting me as I floated, as if in a descending plane through dense fog...

Then the Earth revealed itself through the foggy layers and I saw...

Twenty-Year-Old Me strutting her stuff on NYU's campus.

I watched her, floating at her side as if I were a ghostly older doppelganger. Quite honestly, I felt like Obi-Wan Kenobi, "force-ghosting" next to Luke. But here, my Padawan was *me*.

I floated with her into the school and down the halls like I was my own phantom haunting my haunted past.

I could tell from the look in her eyes that she was heading to her favorite class: *Universal Mystics: Unveiling the Cosmic Consciousness.* Professor Kyle, tall, dark, and irresistibly intellectual, stood as if waiting for her at the door.

She brushed by him, knowing the effect she had. Seeing it in the third person was an ego-boosting mindfuck. I saw them lock eyes—that secret look we had that always led to...

The memory swirled and suddenly all three of us were in his office. Twenty-Year-Old Me splayed across his desk, legs spread as wide as her flexible self could open them as he fucked her with wild abandon.

"Harder! Fuck me harder, Daddy!"

This was better than porn. But at the same time, I felt the hollowness in her. Hollowness that could never be filled by any man.

The memory swirled again and I was force-ghosting next to Twenty-Year-Old Me as she blundered down the cold and flu aisle at Walmart. She was bleary-eyed, obviously in desperate need of anything and everything down this aisle.

I saw her look up suddenly, as if sensing—

Professor Kyle walking toward her. I saw her eyes light up...

Oh my word, I remembered this moment! It was a surreal feeling remembering it as I watched it unfold, revealing itself from where I buried it.

This was when I dared to think we could be something more than just secret fuck friends. What if this was a sign he was sent to help me in my weakened state? What if...

Then I saw his face as he finally noticed her...me?

Wow, this was such a trip.

A painful one as I registered the feeling his eyes had given this version of me. He glared her way and made the slightest, almost imperceptible head shake as he brushed past her to...

His wife. Gorgeous and glamorous with what could only be their beautiful baby. The infant coddled in her arms was red, sweaty, and fatigued with a similar sort of cold that Twenty-Year-Old Me must have had.

I watched them walk off together, floating next to my twenty-year-old self.

Professor Kyle didn't even look back at her.

She responded with a roll of her eyes and a casual shrug—she was Samantha Sands and had better things to do than mourn a man who cheated on his wife, who never really valued her...

But in this frozen moment alone with this self, I felt the knife-twisting pain of betrayal—shame—disgust—humiliation. All the self-judgment she'd locked away beneath a dam of pride, I now let burst within us both.

It all came crashing down.

A tidal wave of malicious misery, rooted in the feeling of worthlessness.

I felt us both drowning in it, suffocating in all that we'd refused to feel, stemming from this moment when she'd erected the unyielding wall in our mind that sealed these awful emotions away from us.

Fuck me, I wanted to do the same again.

Just let her push it away, block it out, and be done with feeling this way.

But I was my adult self now, and despite still being a bit fucked up, I knew I had to feel this to heal it, to see beyond this, to accept—

ACCEPT!

I felt the truest meaning of that word resound within me, flooding my being like a balm of cool ocean water on the lava that these feelings were.

And in that calming caress, Twenty-Year-Old Me locked eyes with me and held my gaze with a serene smile as she dissolved into me like seafoam.

I felt her.

I accepted her.

I loved her.

And with that love the tides turned in my heart as the memory swirled around me, dissolving into another...

I was back to force-ghosting, but now I was next to a fourteen-year-old version of myself, bent over a red Porsche, clenching a hideous spoiler while getting fucked from behind by...

I had to move around to see—ugh, Trip Lawson, jock-type, party-master, douchebag.

"Fuck me harder, Daddy!"

I froze hearing this come from such a younger version of me. That hollowness I'd felt before now felt like black ice shredding away my insides.

I knelt down near Fourteen-Year-Old Me, as if to soothe her, as if to comfort her. That's what she really wanted.

That's what we've always wanted.

Her vacant eyes shifted and looked right at me. Sorrow slashed my heart as the memory swirled to...

Glenn Heights Senior Private School. Home of the Glenn Heights Golds. Its logo: a football crossed with gold bars was worn with the worst kind of pride on Trip's letterman jacket as he swaggered down the hall.

I was force-ghosting next to Fourteen-Year-Old Me as she peeked out from behind her locker. That similar light Twenty-Year-Old Me had in her eyes when she saw Professor Kyle in the cold meds aisle sparked in hers now as he made eye contact with her. I saw the hope, the yearning, even though she knew his type and reputation.

I remembered then the thoughts behind this yearning...

What if I'd changed him? What if fucking me woke him up to see that I'm worth being with...?

Then that yearning was dashed as he smirked and shifted his eyes to Stacey Klein. Dressed more like she was going to fashion week, this girl owned the school and very obviously knew it. Her gaggle of cliquey cunts surrounded her, clamoring for a piece of the spotlight. Trip sidled up and draped his arm around her.

I felt the crushing blow of it all emanating from my teenage self. But as an adult, I now realized it had nothing to do with this douchebag guy that I'd never really wanted. It was the feeling I was chasing.

The feeling of being chosen.

Trip whispered in Stacey's ear and they both flicked looks of insidious amusement at Fourteen-Year-Old Me. Stacey whispered to one of her cunts-in-waiting, starting a nasty game of telephone that ended in an uproar of laughs and jeers directed at my teenage self.

Suddenly, the hallway spiraled open to a vortex of memories coming at her like a kaleidoscope with a vengeance. Each segment a fractal of Stacey, Trip, and their crew laughing, pointing, gossiping, and threatening. Whispers and cruel laughter echoed all around us...

"Slut!"

"Loser!"

"She called me daddy when I fucked her, eh?"

"What a freak!"

"No wonder her mom offed herself."

"Hahahahahahahaha!!"

My teenage self clapped her hands around her head and crumpled to the ground. The urge to rise up and murder every last one of those motherfucking bullies rose up in me like an inferno.

I felt the heat of it scorch my insides and I doubled over in pain.

The intensity of it brought me to a precipice from which I could see beyond my pain—into the pain of those who were hurting me.

I looked into Stacey, seeing flashes and flickers of her father molesting her, her mother turning a blind eye, jealousy palpable between them.

I shifted my gaze to Trip, seeing flashes of him fumbling on the football field, his father backhanding him in private afterwards, disappointment exuding from him like paternal poison.

I felt their pain as if through a membranous bubble, leaving me unaffected yet informed.

A beatific grace flooded through me as I turned to Fourteen-Year-Old Me and hovered my hands over her head. An augmented bible verse I'd never been able to agree with, yet now knew the power of, flowed from my lips unbidden.

"Forgive them, Samsey, for they know not what they do."

She cringed beneath my hands, clinging to her hate with self-righteous indulgence. I felt that hate within me now, a suppressed storm that would never cease. Not until she...

"Let it go, Samsey. You're only hurting yourself."

I flooded her with the feeling of acceptance and forgiveness, feeling the energies flow from me to her and back in an infinite loop of self-healing.

Suddenly, she looked up at me. We locked eyes in the purest moment of love I'd felt up until that point. She smiled as she pixelated into me like a digital double shattering and coming together into the home of who I had become.

I felt...

Accepted...

And loved her.

The feeling of coming home to myself swirled within me as the memory dissolved once again...

I knew where it was taking me.

I'd known all along.

A cold, clinical hospital room materialized around me. A ten-year-old version of myself had just snuck in the door to find...

Our dad, Jasper Sands, comatose in a hospital bed. A creepy robotic sphere buzzed around his face; a laser-like implement extending from it was working away. Ten-Year-Old Me couldn't see what it was doing, but I now could.

I stared in amazement as the bot shone a laser on a huge, disfiguring wound slashed across the right side of my dad's face. I'd never seen this damaged side as a child, and gasped seeing how wounded he'd really been. The shocking pain morphed into befuddled awe as I watched the laser knit my dad's face back together.

The awful gash that had split his face was no longer there. Not even a scar remained.

It had been called the most advanced AI facial reconstruction to date at the time, but I'd never allowed myself to understand the miracle that had been performed on him. I realized then how much I'd let anger, grief, and sadness drive me from this moment on.

As if to punctuate my realization, Ten-Year-Old Me screamed, "Get away from my daddy!" as she ran over and jumped up on the bed, swatting the thing away. It bleep-blooped indignantly as it crashed to the floor. She glared at it with a vile hatred no ten-year-old should have even an inkling of.

"Don't worry, Daddy, I stopped the evil droid with the Force. You're safe now." She turned to him with an eager smile that quickly faded when she leaned over him to look in his eyes. Eyes that were vividly vacant. As if something not quite him was staring back at her.

"Daddy?"

Nothing.

I saw as much as I felt the desperation in her rise like heat in an about-to-burst bubble. "I'm sorry you got hurt on your way to get me. I'll stay in school all day from now on. We shift like sand, right?" Still nothing. She put her tiny hands on his chest, as if to nudge him awake from sleep as she'd always done when she had nightmares and needed to sleep with him. "Daddy!? Please wake up. Daddy!—"

Suddenly, his mouth wrenched open—a mimicry of human speech spewed forth.

"Hello, my name is Jasper Sands, CEO of Sands Studios, brother...

widower...father?"

With each word he sounded more and more like a robot who's battery was dying. Which, quite literally, he was.

Ten-Year-Old Me didn't understand this.

Instead, she screamed.

I burst into tears seeing her agony, feeling it tear through me like a cold, rusty knife hacking away at my heart.

"Daddy?! Daddy, it's me. It's Samsey. What's wrong? What can I do?"

She threw her body on his, as if suffusing him in love would bring him back. She sobbed into his stiff shoulder as the garbled robot voice droned on. "Sam - - Sam—tha - - Samantha!"

She jumped up at that, not noticing the door open behind her. My memories remembered the conversation between the doctor and Uncle Jeff as they walked in.

"...it's very early stages of this AI but your brother's will specifically states that we try it."

"Try what? Going against the Divine Plan?"

Both stopped dead as they saw Ten-Year-Old Me. It was surreal seeing this constructed version of what I know happened. As the ten-year-old experiencing it had eyes only for...

Daddy.

And those lifeless, terrifying eyes. They opened, wide, and empty as his jaw clamped shut, and he spoke no more.

Ten-Year-Old Me screamed as rage overtook her. She thumped her fists on Dad's chest, slapped him, and screamed in his ashen face. "Come back, Daddy! Come back! Stay with me. PLEASE!!!"

Orderlies came in to pry her off him. Sedation was needed. Oh, so very much needed.

I let those feelings I'd refused to feel wash over me...

And all went dark.

CHAPTER 6

I blinked awake on my office couch, staring straight at an art print of M.C. Escher's *Relativity*. The series of crisscrossing staircases in a labyrinthine interior had always soothed rather than confused me.

While the piece was built around the famous Penrose Stairs, AKA "the impossible staircase" that ascends and descends in a connected, continuous loop, I'd always seen what was possible. As Jeff Goldblum famously said in that retro dino film, "Life finds a way..."

Suddenly, a glass of water was in my face. I jumped and turned to see Brad, offering said water along with some much needed Tylenol.

I scanned my disaster of an office. Desk strewn with booze, supplies all over the floor.

Well, shit.

I accepted the goods from Brad, downing the water in one go.

I looked up at him nonchalantly, "So...IRIS and I had a great chat."

Brad's too-charming-for-his-own-good self broke out into a rich baritone laugh.

It caught on, and fuck me did I need it.

***

After I'd freshened myself up enough to not smell like a gin factory, I sat with IRIS, Brad at my side, scrolling through my input.

"I don't remember typing any of this."

"You may not have typed all of it, but IRIS responds to both typed input and voice commands."

"Mmmhmm," I responded, half listening as I read the words that described the memories I'd just re-experienced. It was a mess, definitely a barf stage kind of draft, but there was a story here...my story. One I wouldn't have unearthed without IRIS.

"I would have reconsidered my approach to your intro to AI had I known about your father's history."

"Oh, that's all in the—" Past? Was it though? It was hard to brush it off with one of my deepest wounds literally staring me in the face.

"Thank you," I acknowledged instead. I felt my usual wall of pretending to be fine rise up. A big part of me wanted it there, but something softer had emerged within me that couldn't breathe behind it.

So I scoffed at myself and let it rip instead.

"It was a seriously fucked-up event, and no joke, reliving it just now was horrendous. But I had to be strong, shift like sand, and never give anyone a reason to doubt me."

I'd literally scrunched in on myself, clenching an angry fist like some kind of feral creature. I felt like I'd always imagined Gollum from *Lord of the Rings* must feel, grasping so tight to some precious notion of superiority that I thought made me so strong.

I caught Brad's eyes like twin beams of blue fire burning into me and seeing everything.

I cleared my throat as I shifted back into what I hoped was a more normal position.

"Anyway—"

"There's no need to hide anything from me, by the way. I don't make promises I can't keep but I can promise you'll get no judgment from me. Your angsty side is actually a part of what makes you so incredible."

I rolled my eyes and scoffed (yeah, I do that a lot). "Thanks but—"

"Do I strike you as someone who'd say something like that if I didn't mean it?"

I actually took a moment here and looked at him. Really looked. Whatever happened to me in that memory regression, self-healing haze that IRIS' questions had spun me into had left me feeling amplified. I was more aware of and in tune with the energy around me. I could feel my Merkaba spinning in a heightened way, and although my lower green pyramid was still not quite aligned, I could still feel a genuine intention of support emanating from Brad. It was laced with fascination, woven with understanding that went even further...

I blinked, realizing I'd gotten lost in my head. Shit! And yet...

Brad still sat there, unchanged since he'd asked the question. It made no sense, I'd been meandering in my mind for a good few minutes. I was sure of it.

Or was I?

"No," I finally said. As I spoke, the all-pervading sense of time itself seemed to speed up again, and align back with me. I felt that reverse lightspeed feeling again, but here my mind was like a ship jumping through it.

The question was, who was controlling it?

I shelved that rabbit-hole-inducing thought for later.

"You strike me as a man of his word. Literally. And I don't use that word lightly."

"Of course you don't."

As he said those words I felt an extreme sense of déjà vu. Time now seemed to expand and I saw a memory before me—overlaid on top of this present moment like tracing paper over a design.

It was a memory of me at a Comic Con panel discussing the use of symbolism and metaphor in fantasy and science fiction to represent complex psychological concepts.

Someone in the audience raised their hand. It had annoyed me. That annoyance had spiraled out in my mind into resentment at this person for making me feel...interrupted?

Ugh, what a bitch I was! But no, I couldn't go down the usual self-condemnation spiral. That was who I used to be but now I could look...

Under the annoyance...

Where this me had felt disrespected...

And under that, I felt what it was really all about...

Fear that I wasn't good enough.

It felt like an inevitable discovery as I realized it. And yet—it seemed to have taken me lifetimes to really see it as I did now. It wasn't easy by any means, and I felt a resentful part of myself wanting to hold onto that not-good-enough feeling. But now that I knew that's what was driving this past version of me, I gave her grace as she nodded to the audience member with a stone-walled smile.

"Yes, you had a question?"

The audience member stood up—it was Brad! The Me in this memory just didn't register him with all the noise in her mind.

"I was curious about the relationship between Syeda and Barathon. Was Syeda's eventual ability to bring Barathon back to life with both her magic and ancient tech a metaphor for her forgiving herself for not being able to save her father?"

Comic Con panelist-me turned up the corners of her mouth—just a touch. She still didn't look him in the eye, but I remembered how the inner angst was quieted by the joy of a fan getting it.

"That's literally the metaphor that sparked my imagining of their entire relationship. And I don't use the word literally lightly."

"Of course you don't." He said it with that same smile and fascination he had for me in the present.

A shiver ran through my heart as I let myself feel and know that he really meant those words.

Then the memory overlay dissolved away...

That lightspeed-effect time realignment shift back to the present happened again, but this time I was prepared for it.

"Holy shit! We've met before."

"Comic Con 2045. Glad I *literally* jogged your memory."

"Ugh, I can't stand how much that word's been made a mockery of itself by imbeciles." I crossed my arms, getting into Gollum mode again. I suddenly felt a vice-like pain around my abdomen and the suffocating feeling of my energy field constricting around me. I gasped, uncrossing my arms. The life-giving breath of air relaxed my body and my energy field expanded.

Huh, neat.

"I've changed a lot in five years," I began, then went into over-explaining mode. "It was my first panel and I wasn't expecting questions. I'm sorry if I–"

"Let's leave the sorries to the Canadians."

I laughed at that. The loud, boisterous kind I don't usually like anyone to witness.

Brad seemed to light up at my crazy chortle. What a strange man.

"Seriously though, I just had the weirdest déjà vu that relates to what I unearthed in this memory regression." I scanned the words on IRIS' screen again, ten-year-old me flying into a sedative-inducing rage while begging her father to come back to life, thinking he didn't because she wasn't good enough to come back to.

Just like so many of my characters.

Instead of shaking it off I let that settle into me. "I honestly thought I was over my dad's passing, but..."

"The shadow of the unconscious hides many secrets," he said, eloquently and effortlessly.

I quirked a brow. "Are you sure you're not a writer yourself?"

"Not a writer of your storytelling caliber, but I write in my own way to create tech such as IRIS." He paused there. I could tell he was formulating his next words carefully. I felt the energy of wanting to protect me emanate from him. A part of me was annoyed at this, but more of me accepted it and dare I say...welcomed it?

"I've also recently been involved in a renaissance of the AI your father insisted be tested on him." He declared, treading oh so carefully.

"Lazarus Online?" I hadn't said those words in years, so afraid of the hungry ghosts that might come up and devour me.

Brad nodded. "Unfortunately, at the time of your father's death, technology wasn't quite capable of resurrection. Now..." He trailed off. I could tell he was lost in the creativity of his mind, as I often was.

"What? Now it is? Isn't that going a bit too far? 'Messing with the Divine Plan' as my Uncle Jeff would say."

"Yes...and no. Conversely, what if the fact that the potential for it exists now is because it's meant to be? Maybe we've evolved enough consciously as a species to utilize it properly."

"Maybe *we* have." I gestured to us. "But there are still so many people in this world who I wouldn't trust within six feet of anything more advanced than a millennial toaster."

"That's actually what held me back from going all the way with the project at first, but something I can't fully explain told me to trust that there are people out there who will use it for the greater good."

"I get that, but people who think of the greater good are few and far between."

"The quality of those who do outweighs the quantity of those who don't."

His words brought to mind an image of a scale:

On one side I saw people haloed in bright light, a small cluster of light was flowing through them and around to everyone in a continuous loop. On the other, there were dark figures, a swarm of them compounding on top of one another like tetris blocks building and destroying again and again.

I glared at the dark side with seething contempt and at the light with remorse and pity.

"Your optimism is inspiring. I hope it doesn't lead you into disappointment." I said to Brad.

"I trust it won't. It led me here after all."

We shared a smile at that. I felt a sense of deep alignment. "Well, I'm starting to see why the universe brought us together. What do you see Lazarus Online doing for the world?"

He took a few moments to ponder, which I appreciated. His eyes lit up like sapphires struck by sunlight as he formed his answer. "I see it bringing hope, possibility, and love back to a world that's so afraid of death they do horrible things to avoid it. Keep in mind it's still not something that's completely possible yet, but between us, my team and I have made significant advances that are evolving exponentially as we speak. Getting IRIS fully functional was a huge part of the momentum behind it all."

Where I once would have been repulsed, I now leaned in. "Has anyone testing IRIS ever had an experience like mine?"

"Definitely not. Your input made this a unique experience for the both of you."

I glanced at IRIS, feeling a strange camaraderie with her now...Her? Yeah, it felt right to call her that now.

"You know, before I heard she was an Intuitive Response... ummm...?"

"Intuitive Response Ideation System," Brad finished for me.

"Yes, that! I actually thought you might have named her after the Greek messenger goddess of rainbows."

Brad raised a brow this time. "You're not far off. That tie-in to her name is where we determined her color scheme." He nodded to the rainbow icon. "But in truth, I named her after my mother, who always told me to trust in the messages I know are from God, no matter what others think—especially those in the tech industry." He rolled his eyes in a way that told me I could be frank as fuck with him.

I smiled, realizing how much I needed that. "Where was this guy when we were arguing?"

"I've always been here."

We locked eyes at that. He had been, hadn't he? I cleared my throat and blinked away before I lost myself in his gaze.

"Well, it's nice to meet you, Brad Falls. I'm Samantha Sands."

"Pleasure's all mine, Miss Sands."

I gave him that look I knew arrested men's attention—but now with the intention of choice, owning what I wanted in the moment, trusting in the best possible outcome for both of us. My sultry smile clinched the deal. "Call me Samantha."

He laughed—a rich, warm rumble that sent a volt of heat to my pussy. I shifted in my seat so my erect nipples wouldn't show through my blouse.

Men seem to have a sixth sense about these things though—freaking pheromones.

He leaned closer, breathing me in. My body couldn't help it and leaned in too...

Suddenly—and when I say sudden I mean lightning fast—I felt that familiar urge to rush into fucking, to give him what I thought he wanted. I went to clench my fist but instead breathed...

And that frantic feeling faded to a steady, beating desire.

Nipples still stabbing through silk, I pecked him on the cheek and whispered in his ear, "I've seen way too many of *The Fast and The Furious* films, but if you're in for something languorous like *Waiting to Exhale*, I get the feeling it'll be worth it.

"Languorous is my love language."

I laughed again, which seemed to have the same effect on him as his laugh had on me.

"Good. I'll need someone of professor level to rein in my rushing."

"The difference between rushing and productivity is desperation, of which you are the opposite."

"Oh, the dichotomous dance we lead."

"Who's leading?"

I smiled at that and looked up to the ceiling, softening my gaze. I saw more than the weird rigid whorls of plaster and concrete; I saw the patterns of the universe flowing just beyond a veil I was starting to realize had always been obscuring my true sight. "Uncle Jeff always called it 'The Creator.' I don't know though, is it any one thing? Is it God? Maybe it's a Goddess? You know, even Yoda floating on a cloud up there wouldn't surprise me. Whoever it is was with me in that memory regression."

Brad looked at me for a good long moment, as if seeing into me and formulating his response.

"What?" I finally asked.

"Well I couldn't help but notice that you quoted scripture to your teenage self in the transcript of your regression. Have you read the Bible?"

That was the most unexpected question I could have gotten at that moment, but at the same time it got me to think back to my memory regression...

"Forgive them, Samsey, for they know not what they do..."

I blinked back to the present, Brad's azure eyes awaiting my response. I cleared my throat before answering, "Not really. I went to church a couple times with my dad as a kid but honestly, it didn't stick. Felt like too many rules. Interesting that's the part that stood out to you though, were you a pastor in another life?"

He paused before answering, something in between a smile and a frown at play on his face. "No, but my father was, and while I wouldn't call myself devout, I do believe Jesus Christ said those very words of forgiveness as he paid for our sins."

Something in me scowled at that while I did my best to maintain my not-so-bitchy resting face. "See, that's a huge part I never understood. What kind of father lets his son be crucified? And moreover, what kind of man could ever forgive those fucks for fucking him over like that? And for what?" The Gollum in me was trying to return and I made a conscious effort to relax the rigidity that had suddenly come over me.

Brad just looked at me. This calm, nonchalant smile on his face, as if he knew something I didn't. I didn't like it, but I was also intrigued by it.

"Forgiveness sets you free, Samantha. That's the lesson to be learned here, and through what you've just experienced, it sounds like that was achieved, no?"

I thought about that, feeling back to when I hovered my hands over my teenage self and spoke those biblical words—something had happened there, a surrender of sorts I couldn't quite yet comprehend.

"To be continued," I said with a wink, filing that away in my ever-expanding mind cabinet. "We're here to focus on IRIS and this AI implementation thing, right?"

Brad smiled, looking like he was doing his own mind filing. "Yes, ma'am. To be frank, this level of depth wasn't what I expected, but considering your storytelling, I'm not surprised." He looked at IRIS, then to me. "Do you want IRIS to lead you to more?"

I glanced at the rose-gold sphere that now looked quite lovely. "I do."

"Why?"

"To create from what she leads me to."

He smiled and got up. Pecked me on the cheek and whispered in my ear as I had to him, "Creation is your love language."

That left me at once breathless and full of vigor.

At the door, Brad turned back. "I'll leave you ladies to it, but when you're ready, I'm taking you out for a night you'll long to regress back to."

And out he walked. A cynical urge to think I'd never see him again because I didn't put out rose up and shattered against a newfound trust in myself I realized I had in that moment.

I turned to my rosy new friend. "OK, IRIS, let's create."

CHAPTER 7

The next morning I had a revised outline for *Digital Dystopia* I'd never have known was in me if not for IRIS.

I was astounded. An outline of this depth and scope usually took me at least a week, even with my head writer writing with me.

I'd felt egoic shadows rise as IRIS seamlessly learned and adapted to the discovery portion of my writing process. Even knowing these were parts of myself that wanted to cling to the angst I'd conditioned myself to feel toward AI, it was hard to let them go. It was a mindfuck to realize that now, somehow, with IRIS by my side, I felt not just supported, but seen.

Good God damn—Had I become a pro-bot bastard?

I laughed as I hurried my steps, hearing them now as a dancy, staccato beat—

CLIP CLOP CLIP CLOP—it felt nice not to be in Boss Bitch Mode anymore. But I knew she was in there, trigger-ready if I really needed her.

I smiled, accepting this about me now. I couldn't wait to tell—

"Uncle Jeff!" My voice met stark silence as I ran into his office.

Something felt...off.

Suddenly, the curtain to the Zen Dome fluttered and let a shaft of sunlight into the room.

"Ready or not, here I come!" I heard in my mind as a memory of Uncle Jeff sprang up...

He'd shout the tell-tale phrase during our regular game of hide-and-seek we'd played throughout my childhood as he pretended to seek me through the office to where he already knew I was.

Where I knew he was now.

He had to be.

I smiled, trusting in that truth. The last time I'd been met with this kind of silence, he'd gone on an Atlantic City bender right before a huge pitch meeting. I'd had to step in for him and I was terrified—

Suddenly, resentment boiled up within me like acid reflux. I was tempted to push it down with the emotional equivalent of way too many Tums, but something in me now knew better, knew I had to let it rise. As I did I felt it peak—it hurt, but with the pain came a profound release.

A letting go, not just of resentment, but of all the expectations I'd projected onto Uncle Jeff since Dad died. After all, I realized in the wave of that instant, it had worked out in my favor, I'd crushed the meeting, getting us the deal from which all the rest sprouted.

And from then on I was never afraid to pitch to anyone.

Everything I experienced in that moment shivered through my body in seconds as I felt my energy return to it. Wherever I went to process what I'd just released was higher than this reality somehow, somewhere...

I filed that mystery away for later as I stepped purposefully towards the fluttering curtain.

It was time for me to step up again, into my own version of adult.

I tore the fabric back with a flourish — "Found you!"

Ha! I was right. Uncle Jeff sat on his meditation chair of self-debauchery, sunglasses on, and his treasure chest laid out beside him.

Every box opened.

Fuck me, it was worse than *Leaving Las Vegas*.

He didn't move.

I scoffed, this was our little game. An inside joke he always used to play on me. *"You took too long and I froze,"* he'd say. *"You have to bring the fun for me to rise from the dead of boredom."*

I snatched the still-smoking cigarette from his mouth and took a long deep drag. I felt the addictive urge in me rise but at the same time I knew I didn't need more as I enjoyed the headrush.

"Really, Uncle Jeff? At least if you're gonna play that joke, set the stage."

He still hadn't moved.

"Oh, come on...come on Uncle Jeff—" my voice cracked but I pressed on.

"These can be my finder's fee." I plucked the Ray-Bans from his brow, posing like those over-the-top models we always made fun of.

I felt the tears stream down my face before I dared to look at him. When I did I saw what a part of me already knew I'd see.

Death.

The finality of that thought tore a wail of grief from my throat. I collapsed to my knees before him, gripping his hands as if to hold his spirit there, with me always. "What the fuck, Uncle Jeff!? You promised! You promised you'd take care of yourself! You promised you'd always—"

My rage had superseded my sorrow enough for me to look him right in his eyes—wide, solid, and staring—looking like marble orbs rather than human eyes. I was about to go into another tirade when I felt the last dregs of that resentment toward him that I'd just released clawing at me as it was falling away. I accepted it as I let it go, and it was gone.

It was then that I saw something flicker within his eyes. Not life exactly but—

I ripped off his sunnies and peered closer.

Nothing. Nada.

Actually, wait...

There, so deep I couldn't actually tell if I was making it up or not, I saw a soft glimmer...or maybe I felt it and the feeling inspired the vision? Whatever it was, I knew there was something of him in there.

"Good morning, Mr. Sands!" Cindy's cheerful voice echoed from within his office. Where I once would have cringed, I now lit up.

"Cindy! Help me!"

She was out in a flash. I saw her take in the sight of my uncle, and the sorrow in her eyes told me all I needed to know. Then she started to cry.

"Hey, Cindy, it's OK. We can help him but we need to stay calm, alright? Can you do that?"

She turned her doe eyes on me, and I felt her energy shift from scared to resolute.

I felt the power of that shift shiver through me too and we exchanged a silent nod. It was more than a yes, it was the start of a bond between us I'd never expected.

"Can you trust me and do exactly what I say?"

She nodded again.

"Go get Brad Falls. Tell him to bring Lazarus Online, and tell no one else."

CHAPTER 8

When Brad arrived I couldn't tell if a second had passed or a whole day. I'd slipped into some kind of energetic timespace where it was just Uncle Jeff and I. Time here was not something I was in, but something I was with, flowing through me and generating a force that I couldn't name by anything else other than love.

There had been no thought, only action, as I'd laid him out on the floor of the Zen Dome on a Flower of Life yoga mat.

I'd placed my hands on his heart and let my energy support him like energetic CPR. I felt a strange sense of dissociation; I was in my body but there was something greater within me, working through me, that knew what it had to do.

That's not to say that other parts of me weren't hysterical with grief, loss, and rage. But I was detached from them in a way that let me accept all the emotions and not give into them.

I felt like I was in that higher place where I'd processed my earlier resentment—where time, space, and energy were all one, as were life and death.

This ending of Uncle Jeff's life was flowing into a new beginning that I got to choose to create.

Something in me knew his body wasn't coming back online. I could feel the irreversible damage to his heart. But on an energetic level, life still flickered within him and somehow, my energy was sustaining him. I shuddered at that thought while also marveling at the power I had. With this power I could do anything. With this power I could defy time itself...

Could I also defy Death itself?

It was then that I heard Brad's voice drone in as if from far away. "Samantha? I'm here. Let me help."

Despite the grace in which most of me was floating in this time-warped bubble, I felt a very real part of myself rise with the urge to lash out, like a lioness protecting her pride.

Instead, I looked up at him. His calm blue gaze grounded me. I nodded and let him do his thing.

He had a silver metallic case with him from which he unpacked wires and tech contraptions I had no name for. I had so many questions, but I held off on analyzing and let trust take its place.

Before I knew it, Brad had Uncle Jeff wired up with electrodes all over his head. He pulled out a high-tech tablet, looking like a space-age wizard as his fingers flew across the screen.

"Samantha," Brad started, not looking up from the tablet, "I'm beginning a process called neural mapping." His fingers continued their dance over the illuminated screen, lines of code flashing by. He nodded to Uncle Jeff's electrode crown. "These are cognisan patches, capturing Jefferson's neural patterns and mapping his connectome."

"What's a connectome?"

"It's like a city map of the mind," he said, staying focused on the task at hand. "Imagine each neuron as a building, and the synapses are the streets linking them. This mapping is capturing the pathways of Jefferson's thoughts and memories, sketching a blueprint of his mental landscape."

He tapped the tablet, illuminating a complex digital network. "See, every dot is a neuron, each line a connection. It's a snapshot of his cognitive universe, a glimpse into the architecture of his consciousness."

He paused, swallowed hard, then added, "What we're attempting here is highly experimental. And while it's not a resurrection in the way of Jesus Christ or Lazarus himself, it will preserve Jefferson's essence digitally."

I let that all hit me in the silence that followed as Brad placed more cognisan patches around where my hands lay on Uncle Jeff's heart.

Could this really work?

The initial inspiration that called me to do this suddenly felt like a stupid, silly, far-fetched notion that spiraled into doubt and sucked me into a debilitating déjà vu. The pain of it cut deep like I'd just tapped a root of rage all sparked from Brad's two simple words...

Lazarus himself.

In the flash of a bygone moment, I saw 10-Year-Old-Me reading the Bible, fixated on a passage where Christ brought this Lazarus back from the dead. I felt her hope, her longing, her trying to believe it would happen for Daddy...but it hadn't.

"Samantha?"

Brad's voice brought me back to the present.

"What?" I snapped, then caught myself, realizing that was my 10-year-old self's anger and it had no place here. "Sorry. I just got taken back to a bad memory." It was then I noticed the tablet flashing red. "What's happening?"

Brad looked from me to the tablet. "Tell me about this memory."

"We don't have time for that, Uncle Jeff—"

"I need to know this to save him, Samantha. Where did you go?"

I felt that rage root get tapped again and the tablet flashed red faster. My eyes widened and I looked at Brad, his cool blue gaze calmed me and I took a breath I didn't realize I was holding in. The tablet's flashing slowed.

"When I said I'd not really read the Bible that was kinda half true..." I took a deep breathe and continued, "When my dad was undergoing the Lazarus Online procedure, I looked up who Lazarus was and went to the source." I suddenly choked up on tears, "I believed—I really did—that it would bring him back—" I felt that grief-stricken rage of my 10-year-old self rise up again. "But it fucking didn't!" The tablet flickered like a raging ambulance siren again.

Brad put it aside and knelt next to me, "Samantha, listen very carefully—don't take your hands off Jefferson—but look at me."

His cadence compelled me to obey and in his eyes I saw a peace that surpassed my understanding. "I can't imagine the grief you endured and it makes sense you'd have lingering doubts, but truly I tell you, your belief in this working for Jefferson is the linchpin on which our whole operation rests. I more than know this, I believe it. Do you trust me?"

I wanted to break eye contact, it honestly made me incredibly uncomfortable, but under that discomfort there was a stronger compulsion to hear him, to trust him, to believe him.

"Jesus himself said it's our belief which powers the works of God, and whatever's going on here is coming from a higher power than us. Do you agree?"

Jesus was a concept I couldn't fathom at that moment, there were so many doubts and blocks there...but Brad? Somehow, a part of me I didn't even realize was there until that moment, trusted in him.

"Yes." It came from my lips automatically as if this higher power he was talking about was working through me.

"Do you believe this can work?" He pressed.

I looked at Uncle Jeff, at the machinery, at the tablet now back to the blue coloured projection of Uncle Jeff's connectome.

"Yes." I felt how much I truly meant it as I said it. Like a resonance that rippled deep within me.

All of a sudden, Uncle Jeff's neural network displayed on the tablet lit up like a Christmas tree, expanding across the whole screen. Brad's eyes went wide as he picked it up, rapidly scanning the data.

"Phenomenal." He looked up at me now with a mixture of awe and fascination. "Hold onto that belief, Samantha." I sensed how much he wanted to explore the energetic process moving through me but he held off and focused on his work, as I focused on mine.

After a flurry of final taps and types, he paused his tablet work and looked me right in the eye. "Samantha, I truly believe that if you believe this will work, it will. But logically speaking, there are no guarantees here. We're literally going into the unknown right now. Do you consent to this?"

I looked into Uncle Jeff's wide, dead eyes. Eyes that had held so much life and love.

I could feel doubts like shadowy tendrils, trying to pull me back into that spiral of disbelief. I breathed in deep, imagining I was breathing in belief...mine...Brad's...the belief in that higher power working through me.

As I did, a spark ignited within me. I didn't know then exactly what it was, and despite a big part of me wanting to figure it all out before proceeding, I kept her at bay and let what sparked me take the wheel.

"Do whatever it takes to bring him back," I declared.

Brad nodded and fell silent again, his gaze never wavering from the tablet, fingers flying as if time was racing against them. I could see the sweat trickling down his forehead as if in slow motion.

The soft beeps and hums of the equipment around us were the only sounds.

I surrendered to the trust I'd manage to cultivate in Brad, closed my eyes, and focused on the rhythm of my breathing.

I felt the love in my heart for Uncle Jeff flow like a river of energy through my arms and spiral into the palms of my hands. In my mind's eye, my Merkaba lit up around me, the upper golden and lower green pyramids spinning in opposite directions—

Faster than light—faster than sound—at the speed of life itself.

I heard Brad's equipment go haywire, but as if from far away. I paid it no mind as I visualized gold energy spiraling into one palm and green energy into the other.

Both palms pulsed, hard and fast; then the energies flowed into Uncle Jeff like glowing threads unspooling. This was a use of my Merkaba I didn't even know existed. The weirdest thing was that somehow I knew it wasn't exactly me doing this. It was something I couldn't yet name or explain working through me, being the life force Uncle Jeff needed.

Suddenly, I felt a draining sensation, as though energy was being siphoned out of me. But I held on, hoping my efforts would buy Brad the time he needed.

In that shift to hope, something opened up inside me, deep within my heart where the life force was flowing from.

I felt a rising tide of energy—

A pressure—

An expansion—

My heart raced, beating faster than my body could handle. I almost thought I would break with the force of it all, but where I once would have contracted in fear, I leaned in and pushed further.

I did break then—
I broke open...

Like a raging waterfall—more life force than I had the capacity to hold flowed from me into Uncle Jeff. Into the room, into Brad, into the Earth itself, and the universe within, around and beyond me. I felt this force graze the edges of something greater than all of it combined, something that was also a part of what was producing it with me...

There was no limit to this energy, this force of love. It was an abundant fuel that existed, not just within me, I realized, but within all of us. A wellspring to be tapped in moments that force us to go beyond what we once thought was possible.

I felt Uncle Jeff's essence ignite beneath my hands. I felt him in a more intimate way than I'd ever felt anyone, his life force flowing freely because of mine.

I basked in this feeling for as long as I could before my physical body was forced to shut down and all went dark.

CHAPTER 9

I stayed in the dark for a while, lost in a haze of exhaustion. I kept trying to get up, but my body wouldn't let me. Brad's piercing blue eyes broke through at times as he checked in on me, fed me, and did all he could to ease the incessant drive I had to push aside my obvious mental and physical needs to attend to all the things I had to do.

After a time I surrendered to what I knew I really needed. To paraphrase one of my favorite Canadian babes, Sarah McLachlan, it was the sweetest of surrenders.

***

I let myself float in a sublime blanket of darkness and marveled at the fact that I used to be so afraid of it, but now, I reveled in it.

As I floated, I caught glimpses of what I can only describe as other worlds, dimensions, and realities. Windows into stories I'd get to one day unravel into shows. I felt the bliss of doing nothing met with the ecstatic anticipation to create.

Then suddenly—or as sudden as things got in that liminal realm— I was swept into a cascade of dreams...

I was a child again, running through the mansion I'd lived in with my dad. This was different from the force-ghosting I'd experienced in my memory regressions, though. Here, I *was* my child self, and all she wanted to do was play.

It was like her consciousness expanded from within, and the adult me was free to simply be in the corners of my mind. I engaged with my mission of the moment—seek Daddy and Uncle Jeff in an epic game of hide-and-seek that I never wanted to end.

Running through those long hallways, into rooms and secret spaces that had been the constructs of my childhood, filled me with a joyful serenity I thought I'd lost.

I could have stayed there forever.

But even ten minutes seems long to a child, never mind forever.

Impatience grew within me.

"Daddy! Uncle Jeff! Come out, come out!" I cried as I found myself in the longest hallway in that grand house.

Doors and windows on either side seemed to stretch to infinity. Curtains fluttered in the breeze from the open windows, dancing with dustmotes in rays of sunlight. I watched the floating motes, spellbound until—

I heard my dad's laugh, a rich rumble—the sound of home.

I ran toward it like a beacon. The curtains blew fiercely now as the wind whipped up. Daddy's laughter was nearly drowned out by it until—

"Found you!" I couldn't see him, but I knew he was there as I reached through the wind and billowing drapery to grasp his hand.

I felt a sublime sense of safety holding onto him as he emerged through golden sunlight and diaphanous drapes. The wind ceased immediately.

Suddenly, I was no longer a child. I was me. Samantha Sands, 36 years young and facing the spirit of her father. Whether this was a dream or even a delirious figment of my infinite imagination—I didn't care. I pulled the man I would always long for close and savored the embrace I'd denied needing for so long.

Nothing was said. Nothing needed to be. We were in an everlasting moment of pure presence that would live in my heart forever.

Despite wanting it to last, when he pulled away I let him, but I couldn't yet let go of his hand.

We stared into each other's eyes. Eyes that were mirrors of each other.

He finally spoke, his voice a soothing tone. "You found me, Samsey."

"Of course I did."

He laughed. "That's my girl. Now where's Uncle Jeff hiding?"

Suddenly, I was a child again. My tiny hand enveloped in my dad's seemingly huge one. I grinned up at him, full of frivolous mischief.

"Ready or not, Uncle Jeff, here I come!"

I squeezed my dad's hand as I let it go, feeling the grief of it at the same time I felt the love. The love was what allowed me to release him as I ran off to find my uncle.

"Uncle Jeff...Uncle Jeff?"

The hall flickered with images I didn't want to see—

Uncle Jeff and I smoking in the Zen Dome—

Uncle Jeff frozen in the stillness of seeming death—

Uncle Jeff laid out on the Flower of Life yoga mat, dead eyes staring up at nothing—

Suddenly, electrodes sprouted all over him, more than I could count, more than could rightly fit on his body—they were consuming him as gold and green energy went haywire all over—

"Samsey? Samsey...I'm here."

That's when the darkness smashed upon me.

***

I woke up with a start as if electrified from the dead, my gasp a resounding sound that ripped from my parched throat.

I was in a room that was not my own. Brad's room. I could tell by the smell, by the essence of him all around me.

"Samantha?"

And there he was, rushing in to tend to me. As he had for...

"How long have I been out?"

"Today's day eight."

"The number of infinity. How appropriate."

I snatched a glass of water from the side table and chugged it, feeling the hydration flood through my body in a heightened way. It felt amazing, exhilarating even—but I had no time to revel in feelings now.

I threw Brad's feather-filled duvet off me and got up—

Oops! That might have been a mistake, I thought, as I wobbled on feet I hadn't walked on in over a week.

Brad steadied me. "Woah, take it easy."

I felt an urge to push him away, and immediately felt guilty about it, so I let him support me as long as I was able to.

"I've had my take-it-easy time, Brad." I noticed the edge to my voice and eased it off. "Thank you for letting me stay. I don't remember it all but I know you were here."

"I've always been here, remember?"

I looked into his eyes and gave him the kindest smile I could muster. He really meant that. A part of me knew it, but not all of me could believe it.

"Where's Uncle Jeff?"

CHAPTER 10

In the depths of Sands Studios was a newly installed space I'd abjectly refused to visit until now.

Brad and I approached the imposing double doors over which hung a steel plaque I'd sneered at more than a time or two.

FALLS TECH - IRIS RESEARCH AND DEVELOPMENT

I read those words at that moment from an entirely different perspective than I had in the past.

Where I once saw a threat, I now saw possibility.

Brad swiped his keycard and the doors slid open with a nearly imperceptible WOOSH!

I stood at the threshold of a dark abyss until...

BUZZ—BUZZ—BUZZ—

Harsh fluorescent lights illuminated a state-of-the-art lab. Despite my remaining reservations and moral questioning, I had to admit it was a breathtaking spectacle of cutting-edge innovation. Its sleek, high-tech aesthetic was like walking onto the set of a futuristic film that stretched out in a seemingly endless expanse.

"Welcome to my Digital Domain."

I laughed as he cast his arm out in over-the-top display mode. He offered another arm for me to take as if we were an old-fashioned couple about to walk a promenade.

After a slight hesitation, I took it, letting him lead the way. "Well, I appreciate the alliteration. Maybe if that was on the door I would have accepted your, oh, I don't know... fifty invites to come down here."

"I understand now why you didn't. Jefferson told me you'd come around eventually. I'm glad I listened."

"I'm sorry. I must have come off as such an asshole!"

"Well, there's an asshole in all of us." He grinned as I snort-laughed.

Oh for fuck's sake, Sam. You seriously sound disgusting.

"Yeah, and half the population doesn't even know how to use theirs right." I let my sarcasm push through the noise in my head as my embarrassing laughter continued, ratcheting up to loud and boisterous. "OK, wow, it feels like I haven't laughed in years. Sorry, but you seem to be some kind of laugh releaser. Bear with me here," I said between my laughs and snorts.

Stop showing this guy how incompetent and unstable you are, Sam. He'll exploit it for all you're worth. Which frankly isn't much.

But Brad's grin only grew at the sound of my weirdness. "Samantha, I literally witnessed you bring your Uncle's spirit back to life. Are you seriously apologizing for laughing?"

In the breath of that moment I realized how much energy I'd been using to hold back what I thought was too weird, loud, wrong, bad, this, or that about me.

And for what?

For once, there was a blessed beat of silence in my head.

I grinned up at Brad. "Well, I know you wouldn't use the word "literally" in vain, so not anymore! Fair warning though, you just unlocked Pandora's box of weird sounds."

"Well, that's a box that never should have been locked in the first place."

My imagination sparked. "Hang on a sec." I whipped out my phone and typed a flurry into my notes. "A quote just came to me for my next great character..."

I kept typing, suddenly overthinking the combination of words.

"And??" Brad prompted. The genuine curiosity in his eyes earned him the extra '?'

I raised a brow and read aloud, "'Unlock your weirdness to release your asshole.'"

Brad guffawed. "That's true on multiple levels."

He offered his arm again as I put away my phone. There was no hesitation this time as I took it and I felt an ease forming between us as the resistance under the noise faded...

It felt like smoke being sifted out of me.

As it left, I leaned a little more on Brad's arm, feeling lighter and eager to hear more from him.

"But in all seriousness, I've studied enough psychology to know when someone's being an asshole and when there's a wound at play. Thank you for letting me see that wound, Samantha."

"Yeah, daddy issues sure screams business partner, eh?"

"I loathe that term. My human sexuality professor phrased it as a "father wound." Similarly, there are mother wounds too—"

I tripped and stumbled as he said the word mother and a breath of ice blew through my heart.

Brad's arm was a solid anchor at my side as he helped steady me. "Are you alright?"

"I'm fine," I almost snapped—eek, there's another asshole reaction that's gotta go.

"Sorry, go on." I felt the smoke of that noise dissipate from me too as I waved my arm out in mock gallantry, hoping he wouldn't see the other wound at play within me.

He smiled in this knowing way he had, like he could more than read me—he could see me.

Parts of myself that either loved or hated that warred within me as we continued our stroll.

"As children we see our parents, or whoever raised us, as God-like figures. We *literally* can't help it." He gave me a side-eye smile that widened my own and began to calm those warring inner facets of me. I laid my other arm on top of his, feeling the connection between us grow stronger, like a bridge through the noise of doubt.

Suddenly, my head came down from the clouds—Brad hadn't noticed my blip out though...

Ah, I realized. I'd gone into that higher place to process something for a second—or however long a second lasted there. What I'd processed was gone before I could even attempt to analyze or hold on to it as time realigned around me...

"So anything our parental figures say or do, and anything that happens to them, impacts us profoundly. And unfortunately, the painful memories override the joyful ones. It's all due to our—"

"Negative bias," I finished in sync with him. We caught eyes. "Don't look so surprised, there's a Psych 101 course going on in my head at all times. That's how my work attracts the top actors and gets them the Emmys they deserve." I nudged him playfully as we made our way further into his domain.

"Not to mention *your* Emmy trifecta," Brad replied, nudging me in return.

My mind's eye flashed to the three golden statues on display in Uncle Jeff's office behind his desk. It felt weird showing them off in mine.

"Yeah, that was a team effort for sure. So tell me more about this human sexuality professor. Sounds like a hot-for-teacher kinda thing."

Brad slowed down, a look of remembered love in his eyes. I suddenly felt something I'd been terrified of feeling, a green-eyed-monster kind of word I still couldn't actually name, even to myself. I let go of his arm, letting him walk a few paces away until he stopped too.

"Frankly, Samantha, I used to be that guy. Similar to the jock, douchebag, party master aficionado character from your past. But Elena…my professor—"

"Yeah, I figured as much." I caught the snark in my voice—

Geez, rein it in, crazy!

"Sorry, go on," I said aloud, as the insecure noise rose up in me.

He smiled knowingly again. I simultaneously did and didn't like it.

"Elena showed me that true sexuality goes beyond love and attraction—even kink."

We caught eyes at that and in a flash I imagined him fucking me on any one of the smooth, steel worktables arrayed around the lab. No one to hear me scream as he pounded into me again and again. I felt myself getting wet as the sexual fantasy spiraled…

Suddenly, a whirring sound approached, and a bot that looked like a space-age Roomba came sweeping down the hallway.

I watched it glide by, remembering how I'd trashed a similar one Uncle Jeff got when I was a teen. That was the anti-bot me. Now I saw that it was a tool, a helper to do things we don't need to be doing.

"Damn, you have bots for everything, eh?"

"Not everything. For tasks that will save humanity the thing we can never get back."

He gestured us forward and I let him walk ahead of me, covertly letting air get to my damp panties. Thank God I wore a skirt today.

Are you seriously getting caught up in sex thoughts when Uncle Jeff's essence is on the line? Keep your freaking priorities straight, Sam!

Still, it was hard not to consider jumping Brad's bones as he walked ahead of me. And it wasn't just his looks, those were a dime a dozen really, it was the connection we were forming around all of this; that bridge I'd felt just a moment ago that went way more than skin deep.

That is until he led me through an assembly line of IRIS bots in progress. Their sphere-like forms were strikingly intricate, more advanced than any computer, yet devoid of humanity. I shivered, feeling that former chill of relentless progress I had yet to shake. Made me yearn even harder for the warmth of Brad's touch—*damn it!*

I sped up to walk beside him as we approached a backroom at the farthest corner of the lab. There was no identifying door plaque here, but I knew that this was where the Lazarus Online project lived.

"You meant time, didn't you? The thing we can never get back?"

Brad smiled as he swiped his keycard. "No use riddling a writer, is there?"

"Not one who's touched a level of timespace she can't even describe yet."

The doors opened with a BEEP and a WOOSH. I felt a pulse of enigmatic energy as we entered the extended space.

Here, the silver walls echoed a quieter, more intense hum of anticipation. Multiple LCD monitors played out cryptic digital dramas on their bright screens, illuminating the sleek surfaces and lending the room an ethereal glow.

My gaze lasered in on the metallic box, central in the room, hooked up to everything like a mechanical brain. I felt like Indiana Jones approaching the Ark of the Covenant as I walked over to Uncle Jeff's digital reliquary, where the rhythm of his being had been stored. My hands remembered the connection, the vibrant link when his life was woven with mine to create the map for his digital resurrection.

"It's remarkable," Brad's baritone voice broke me out of my reverie. His eyes were intense and reverent as he looked deep into mine, reflecting the dynamic lights of the data streams. "The energy you lent to the procedure...I've never seen anything so miraculous."

He gestured to the box and the digital dance of data projecting all around us. "Your energy did more than just preserve Jefferson's essence; it amplified a network of connections I've never seen before, interlinking his cognitive and emotional responses. It's as if you've brought to light a sort of heart connectome, adding dimensions to his digital blueprint I'd never anticipated."

His gaze shifted from the box to me, a mix of awe and uncertainty. "But now, we have a new Everest to climb—successfully transplanting that enriched essence into an AI unit. Your unique energy has opened doors we didn't even know existed, but it's created an unpredictable beast."

He paused, collecting his thoughts, every breath heightening my anticipation. "This is uncharted territory, Samantha. We're about to blur so many lines in this operation. But," he added, his voice filled with steadfast resolve, "I'm with you in this, whatever it takes."

His words lit me up like a beacon in the storm of our shared uncertainty. It was here, in the heart of the technology I'd once abhorred, coupled with the energy of consciousness and the divine unknown, that we were poised to push the boundaries of what it meant to be alive.

That aliveness sizzled through me and my body responded to Brad's words before my mind could hold it back.

I grabbed his perfectly pressed shirt and pulled him into a kiss.

Now, I've written about kisses that ignite passion in so many scripts, but this was passion in action—an ignition I felt in the very depths of myself.

I felt Brad's passion light up within him, rising to respond to mine. I pulled him closer and suddenly found myself backed up against the smooth, steel table. I bit his lower lip then kissed up his neck and nibbled on his ear.

"Mmmmm," Brad moaned, running his fingers delicately down my back as he kissed me HARD—his tongue a rough and delicious partner with mine. I urged him on, devouring his lips in a frenzied fervor when he suddenly cradled my ass in his hands as he lifted me onto the table as if I were weightless.

He grabbed the back of my neck and leveled me with a hooded

gaze. I smiled in my sultry, seductive way and tried to get us back to making out but he held me back, gripping my delicate neck in his soft, strong hand. His eyes pierced mine and, for a moment, held me in a stare that probed me to my core.

I blinked away under the guise of being playful and started unbuttoning his shirt.

"I want it hard, fast, and rough. Still with me, Professor?"

He shrugged out of his blazer then suddenly, my back was flat against the table as he held me down. "There's a time and place for fast, Samantha, and this is not it." He looked deep into my eyes again and even though I felt the urge to look away, I didn't. "Tell me what you need." One of his hands pinned my shoulder down, the other enveloped my waist. For the sake of the part of me that needed it, I let myself be suffused in the feeling of being small and delicate.

"Rough like this?" He asked. He was edging into this but I could feel the kink beneath the careful. Like a caged lion ready to be released.

I moved his hand from my shoulder to my throat. "Like this."

He tried to lock me into that stare again as he squeezed my throat.

"Yes!" I breathed out as I writhed wantonly beneath him. "Now fuck me." I reached for his pants but he held both my hands in one of his, keeping the other held firm to my throat.

"Look at me, Samantha."

I realized then how much I was avoiding eye contact—how much

I'd always avoided that.

Such a simple thing, and yet...

I looked at him, really looked him in the eyes...

And I saw in them what Shakespeare meant about our eyes being windows to our soul. I couldn't just see his, I could feel it.

An energy rising to meet its likeness in me.

"There you are," he said as he released my hands, trailing his fingers down my inner arm, face, neck, breasts, and stomach in a slow, scintillating exploration. My body responded with shivers while my mind started churning with insecurity.

"I'm going to teach you a lesson in my love language..." His fingers played delicately at the hem of my skirt, inching up underneath to where I wanted him. "Do you remember what it is?"

I cried out as he finally touched where I was heating up beneath my damp lace thong that was begging to be ripped off. Under the lace he probed just outside my slick entrance... softly, slowly, a rhythm I wasn't used to.

What's with this dude? Let's get to the fucking so we can get back to work!

As that thought rang through me I had a sudden realization that it wasn't truly mine. I didn't want to rush this either. Under that noise I wanted to let him explore me, I wanted to be comfortable taking it slow...

Brad had stayed busy as I played my mental gymnastics.

What had he asked me again?

He grabbed my hand and locked eyes with me again as his fingers entered me.

I gasped and squeezed his hand in return. Finally able to hold his gaze. "Your love language is languorous." The feeling of that word reverberated through me, softening where I usually shielded.

Brad must have felt that too as he climbed on top of me. Feeling all of me with his hands, tasting all me with his tongue.

The digital data on the screens jumped and danced, illuminating our bodies as we shed the rest of our clothes and relished in the human aliveness that no machine could ever compete with.

CHAPTER 11

The months that followed were a blur of me holding it all together while Brad and I worked secretly on our Lazarus Online project, and I dealt with the aftermath of Uncle Jeff's passing as it pertained to the studio.

The motherfucker had named me CEO in his will. Much to the delight of the Board.

They all came to his funeral, though. In fact, it seemed like everyone who'd ever worked at Sands Studios was there. Seeing the masses rally to honor him on that dreaded day gave me something to aspire to.

But as I walked through the grand hallowed church on that crisp autumn day, nodding and smiling to the sea of grieving people, I felt their sorrow piling on top of my own. It was like being buried in an avalanche of emotion. I bit down hard on my inner lower lip, tasting the metallic sting of my own blood as I fought to stay upright. I would have pressed on regardless, but at that moment Brad stepped in beside me.

"May I escort you to the podium, Miss Sands?" He asked, offering his arm for support.

I took it on instinct, letting him steady me. I felt myself leaning into him when I noticed the eyes of others on us and stepped away instead.

"Thanks, Professor Falls. Just a little light-headed today. I'm fine though. I've got this."

"I know you do," he said as he nodded and stepped into a pew. He meant it, but I could feel the hurt beneath his words—his want to help me that I'd pushed back at him.

A part of me I was just getting to know wanted to soothe his hurt, ease his pain, nurture and assure him, but it was too much on top of everything else.

I summoned my energy as Uncle Jeff had taught me, letting my Merkaba spin and amplify my aura, shielding me from the emotional onslaught. Strangely, I could still feel it all, but as if through a membrane that translated everyone's grief into the love they all had for my Uncle that was under it.

Ah—much better.

I was almost at the podium where I'd deliver my extremely well-written, and many times rehearsed eulogy when my head writer, Michael Strassner, emerged through the throng. His perfectly tailored Gucci suit and debonair trench coat made me smile as he took my hand and swept into a bow worthy of a Shakespearean actor.

"Oh Sam," he breathed as he straightened up and arrested me with his storm-gray eyes. "May I?" He indicated a hug, knowing to ask. That alone softened me and I pulled him into the most heartfelt embrace we've ever had.

I caught Brad's eye over Michael's shoulder. He didn't mean to stare, and I could tell it wasn't with jealousy, as I'd always assumed relationships would lead to. It was with an understanding that Michael could provide the comfort he couldn't at that moment. A huge part of me longed to rush over, kiss him in front of everyone, and make our "us-ness" known, but...

"Jefferson's in a better place now," Michael murmured as we broke apart. "Like the heaven we built in the Ascendants series, God chose him to go for a reason."

I scoffed, "Yeah, well maybe he needs to be knocked off his fucking cloud." I laughed and simultaneously choked up on a lump of way too much emotion as I remembered the arguments we'd get into over "The Creator" and the supposed "Divine Plan" Uncle Jeff was always on about. The lump won as I remembered my uncle's situation wasn't technically what Michael was thinking. I got glassy eyed wishing I could tell him everything, but...

"Seriously though, I know he's in a better place. Thanks Michael." I squeezed his hand and stepped away quickly before he could say any more.

The ascent to the podium felt like the infinite staircase in my art print of Escher's *Relativity*. My mind was going a million miles a minute, rehearsing my perfectly prepared speech, feeling the weight of the attendees' expectations compounding upon my own.

I reached the top and took my place. My hands shook with the force of all the emotional energy on fire inside me as I took out my cue cards. The same cards I storyboarded and planned so much out with. The shaking intensified to the point where I could no longer even see the words I'd written. My mind was too overloaded to remember any of what I'd rehearsed.

Then I burst into tears.

I knew it was useless to resist so I let them flow. I saw Brad, Cindy, Michael, and, strangely enough, Chairman Ross rise to come to my aid.

With a gesture and a smile I held them off, and picked up the mic instead.

"Thanks, you guys, but I'm fine. I'm... more than fine really. I'm at peace. And you all should be too. Uncle Jeff wouldn't want us to cry over him. In fact, he wouldn't want anything else other than a party. So I say we give it to him."

With that I threw my cards in the air like confetti, laid my phone next to the mic and let it play Uncle Jeff's favorite song.

What followed was an epic church dance party to *Light My Fire* by The Doors. It was so bizarrely out of the norm, yet felt perfectly aligned at the same time.

There was one person, however, who definitely didn't share in the cheer—Chairman Ross.

Through the dancing crowd alive with joy, I saw him stomp off. I rolled my eyes, about to turn back to the merriment I'd created when I saw him stop at the door and turn back to watch us. He paused there for a long moment, and in it, time stretched wide open like a chasm between us.

I felt a strange longing from him, something pure and loving tarnished with regret and shame. I tried to focus my energy to feel it out when his eyes caught mine—in a split-second about face, they hardened up with the force of a padlocked door and he stormed out.

Ouch.

That really would have stung a previous version of me, but in that moment I wasn't going to let anyone else's misery kill this celebration of Uncle Jeff's life. A life that could very well not be over yet.

I smiled at the thought of reintroducing everyone here to an AI version of him. It would be weird, but it would also be wondrous.

I let time fall back into alignment around me and danced it out with those who made our studio not just a company, but a home. We laughed, we cried, and we carried it on to the wake.

***

The next day was another story however, as I stood, hungover as hell at Uncle Jeff's grave. Being here was an energy minefield, and in my depleted state, I couldn't hold my Merkaba in alignment as well as usual. I felt the restlessness of the spirits around here who hadn't fully left this world—it was unnerving and eerie, yet I found myself jotting down notes to myself with ideas for yet another series.

"There's a story behind every door, eh, Uncle Jeff?" I said to his pristine marble headstone. Even though I knew his digital reliquary held his essence, something about seeing his name engraved in stone made me feel more connected to him.

Or maybe it was seeing it alongside the weathered gravestone it sat next to. The one that read:

JASPER SANDS

Father–Brother–Widower
Gentle Giant

I'd come here before filled with so much angst, yet now when I looked at my dad's grave I knew he was out there somewhere in those liminal dimensions I'd glimpsed in my near-comatose visions.

I looked up to the stormy skies and half wished I could join him.

That's when the rain fell.

I smiled, letting it wash over me, and as I did a flood of imaginings filtered through me. I sat under the shelter of the huge oak tree I'd insisted be planted here when we'd buried Dad, pulled out my trusty notebook and even trustier flask of gin and tonic, and wrote until the sun went down.

I wrote about Dad and Uncle Jeff, even my mom's fragmentary ghost made an appearance on the rain-splattered pages. The form changed from stream of consciousness to prose to poetry as I translated the experience this newfound energetic ability of mine truly felt like.

In the end, many of the tangents I scribbled myself into led to Brad. My feelings for him, our connection, the way I both wanted and resisted him. The frustration in knowing there was a block there for me to heal, but not yet being able to.

Did I even really want it gone?

In the end, I ran out of page space and took that as a sign to be still. The tree at my back gave me strength, one message ringing loud and clear through me:

You are enough.

Still, I didn't feel like it...

There was always more to fix.

That thought opened up a dark spiral within me, one I filled with the contents of my tried and true flask.

CHAPTER 12

I found myself buzzed on Brad's doorstep twenty minutes later and spared no time jumping his bones as soon as he opened the door.

"This is the time for fast, Professor."

He didn't argue as we stripped each other on the way to his bed where he pounded me into soft, silken sheets.

Every thrust fed me with a fervor for more.

"Fuck me harder!" I grabbed his ass and urged him on. "Fuck me 'til I forget!"

He thrust into me then—HARD—restrained my hands above my head and slid out agonizingly slow.

"Look at me." His voice was captivatingly commanding. I'd come to enjoy the depth eye contact brought to our sex life over these tumultuous months, but my mind was too full of liquid confusion to hold his gaze, so I went into play mode and nipped at his fingers instead.

"Make me."

He grabbed my throat, tilting my head up. I locked eyes with him and managed to breathe out...

"Harder."

He squeezed—ever so slightly harder—on my windpipe, THRUST into me to the hilt, and released my throat as I let out a scream of wild ecstasy.

"More! Fuck me harder, Daddy!"

I froze as the word Daddy seemed to echo around the room. With it came a reverb of shame. My booze buzz latched onto that like a leech to a fresh-cut wound.

"I'm sorry. That just came out. Ugh, I'm such a fucking mess right now."

He let go of my throat as I turned my face away, ready for him to get out of me so I could get the hell out of there. Instead, I felt his fingers on my face, gently turning me to face him.

"All I'm going to ask from you right now is to look at me, Samantha."

I opened my eyes to meet his—

Twin pools of azure greeted me. I felt an instant injection of security, grounding me in the here and now. I shifted my focus, seeing not just his eyes but the energy of him, like another body outlining his own. In all that I saw I felt nothing but truth from him. A truth I could trust in if I dared.

"What do you see?" I asked as I moved my hips, his erection still going strong inside me.

"Wisdom, beauty, and radical resilience." He replied, meeting my movements like a well-trained dance partner. "What do you see?"

Now there was a question, and it was getting harder to think with his cock capturing my cunt again and again. So I didn't think and spoke instead. "Compassion, strength... ahhh!" He'd thrust into me hard again, the way he knew I liked it. Our bodies had come back into their tantric tango of their own accord it seemed—and I was all for it. I moved his hand back to my throat. "I see *you*, Professor, now teach your girl a lesson." I was soaking wet by now and red hot with anticipation when suddenly, he slid out of me completely. Then, just like that, my hands were tied up in the restraints I'd brought over months ago.

"As you wish, but I'm not fucking you 'til you forget," he breathed into my ear in a voice that almost made me gush then and there. I writhed in my restraints as he leaned over and picked something up.

CLICK — BUZZZZZ

"I'm going to fuck you till you remember yourself."

He brandished our vibrator and got to work.

***

I don't know how long we went at it, but we finally came up for air, revived and ravenous.

I watched Brad order a feast for us on his phone, noticing how his right eyebrow arched as he scrolled and contemplated. My heart swelled like a helium balloon and felt as if it were soaring up within me when...

Yeah right. That's just what it felt like with the last dickhead, too.

And just like that, my balloon burst.

Brad looked up at me as I looked away, tears stinging my eyes.

"Hey." He leaned over and held my hand. "What just happened?"

I summoned my well-honed skill of holding back tears and faced him. Maybe I could have been an actress after all.

"Oh, just the usual whirlwind of everything and nothing, all at once," I said, my voice a mix of sarcasm and sincerity. "Hey, can you actually add a bottle of gin and some tonic to our feast? If you don't have cucumbers or limes I can do without." Now that was completely sincere, and the thought of booze on the way sparked my playful side enough to push away the tears and try my hand at leaning in for a kiss.

Brad just stared back at my sultriest of smiles that had always worked before, but now, instead of kissing me back he caressed my face, his thumb softly grazing my lower lip. "Samantha, how much have you already had to drink today?"

Well, that felt like an ice bucket had been dumped on me. I scooted back from him immediately. Suddenly feeling too exposed, too ashamed, too much.

"I had some hair of the dog at the cemetery earlier. What does that matter?"

"I'm not judging you. I'm concerned. I get how stressful everything has been for you, but didn't Jefferson's OD teach you anything?"

"I'm not like Uncle Jeff, OK? I don't touch the white stuff anymore."

"That's not the point and you know it. Look, while I doubt there's anything that would really keep me from having sex with you, I don't want to just fuck someone who's inebriated enough for her to let me. You're more than that, Samantha." I looked away and he moved his hand to my shoulder, nudging me to turn back and face him. Only when I looked him in the eye did he continue...

"*We're* more than that."

I got quiet for a hot minute as a ton of my noise came up—

Great job Sam, he's finally seen you for the trainwreck you are —

> *He's about to run for the hills, so you might as well beat him to it—*

>> *Did you ever really think this could last, anyway?*

He's just like all the others...

The harsh meanness within me ping-ponged around in my brain that was losing its buzz and transitioning to burnt out.

"I get it, Brad," was all I could muster through it all as I got up. "I'll see myself out."

"Woah, I don't want you to leave. I want to work on this."

He means change you into the perfect little woman for him. Barf, Sam, let's get out of here!

So who was I going to listen to?

I felt it then. The difference between my noise and my voice—like a live wire or a calm stream, angst or peace, lies or truth…

I chose powerfully as I slid back into bed with him.

In that moment of choice, my mind's eye flashed to a childhood memory. The joy it had given me must have shown on my face as Brad pulled me close and kissed my forehead, "What's going on in that incredible mind of yours?"

"Do you remember that Disney movie *Dumbo*? The super dated pre-high-tech cartoon version?"

"It's one of my mother's favorites," he replied.

I choked up as he beamed at the thought of his mother and cleared my throat before I bawled my eyes out. "Your mother has great taste! So you remember when Dumbo gets his magic feather and flies?" Brad nodded. "Well… liquid courage is kinda like my version of a magic feather." I smiled, masking the sadness that accompanied this realization.

"Not many people would be strong enough to admit that, you know?" He wrapped his arms around me, and I settled into his reassuring embrace. "But do you remember what happened at the end of the movie?"

"Of course, he…" I paused, thinking back, seeing the movie's climax play out in my mind in a muted black and white tone. "He's in the circus about to perform, he jumps from a huge burning building, plummets to the ground and his feather is ripped from his grip." I grasped Brad's hand instinctively. "I always closed my eyes at that part actually, but then his mouse friend..."

Suddenly, the movie playing out in my mind transformed into technicolor. "Timothy! Now I remember! Timothy tells him the magic feather was just a gag, that he can fly without it. Then he didn't just fly, he flew higher than ever before on his own." A couple of tears fell on Brad's arm. He kissed the top of my head and held me closer.

"I whooped out loud the first time I saw that," he said.

I saw myself as a very young child watching the movie for the first time, the scene expanded and I saw that someone else was watching with Young Me. I couldn't see them clearly, but they had coaxed her out of covering her eyes to see Dumbo fly without his feather. The light in Young Me's bright green eyes radiated a doubled sense of that joy I'd first felt as I remembered this whole thing.

I laughed, turned in Brad's arms, and kissed him. "Thank you for that. I feel like I remembered a piece of myself."

"Always." He stroked my cheek. "And you know what?"

"What?"

"I guarantee you'll fly even higher without your false magic feather." He squeezed my hand. "I'd even stake your life on it."

"It kinda already is at stake in a way. Ugh, but fuck me, this is so not the time I'd have thought I'd do a sober streak. Feels like I'm taking away my feather too early, to be honest."

"It'll never be the right time to take away a vice from your ego, but this overwhelm and what you'd mentioned about feeling like you're in a whirlwind of everything and nothing—I actually understand that," he replied as he enveloped me in his arms again, creating a sanctuary that felt both safe and unnervingly intimate. I let myself relax, but only to a certain point. "You know, 'everything and nothing' is more than a feeling. It's a fundamental aspect of quantum theory."

My body softened further into his as my mind latched onto his words. "Alright then, teach me another lesson, Professor."

Brad's rich, rumbling laugh felt like a much-needed massage against my back. As he began our "lesson," his voice took on a softer tone, filled with passion that resonated deep within me. "In quantum physics, particles can exist in a state of superposition, meaning they're in all possible states at once until observed. It's like they're everything and nothing simultaneously."

"That's... quite literally how I've been feeling," I admitted, the science nerd in me stirred by the concept.

"It's not too different from how we experience emotions," he added, interlacing his fingers with mine. "Think of your imagination as a hub of superposition, everything and nothing always happening all at once within it. Then what you focus on in your imagination becomes the emotion you experience in life."

I smiled, appreciating the metaphor, and felt my imagination buzzing, as if it knew we were talking about it. "That's actually pretty similar to my writing process, too. It took me years to see it, but everything I've ever written stemmed from an experience, rooted in an emotion that I chose to explore and unravel on the page."

"Exactly. Your choices made your creations possible, and look at your results!"

I took a moment to really take in what Brad was saying, and the image of the three Emmys I'd won over the years came to mind. There they were, collecting dust in my office closet. I couldn't bear to look at them since Uncle Jeff's "passing."

"Yeah, my results are... awesome, I guess, but they've never really been my focus. I write because I can't not."

"And now you have the added benefit, but also the responsibility, of this energetic ability, and it's diluted when clouded by alcohol. Imagine its potency when honed and directed."

I felt resistance rise at that for sure, but more of me knew the truth his words held. "But this energy, this...power. What if I can't get it right? What if I fuck it all up?"

He was looking at me in that knowing way that made me feel even more naked than I actually was. "You'll never really be able to see your own greatness, Samantha. I've come to the conclusion that, again, similar to particles, we need others to see it in us and reflect it back."

Suddenly, tears sprang to my eyes. I pushed through the urge to hide them and smiled as they flowed. "Uncle Jeff was always that mirror for me."

"And he will be again. We've got the power of both our intentions backing us, after all." He squeezed me at the small of my waist, the slight pressure sparking pinwheels of realization within me.

"I always used to think that sounded too easy. You know, 'Just intend for the best and it will happen!' Yeah right—but actually... that is right, just not in the way I always saw it, you know?"

"Well, you've had several major quantum shifts lately, so it definitely wouldn't look the same from your current perspective." His eyes lit up. "It's the same principle as with observing a single state from superposition. Our consciousness and intentions— they're more powerful than most people realize. They shape not only how we perceive the world but how we interact with it."

Whatever spark lit him up suddenly lit me up, too. "It's like the Merkaba!" I exclaimed. "And our energetic field. It's all connected, isn't it? Our thoughts, feelings, the energy we put out..."

Suddenly, the lit-up feeling I'd just had fizzled out and the beginnings of a hangover I'd have loved to drink through started to set in. I rubbed my temples and chugged some water.

Brad watched me, weighing his next words carefully. "Your energy in particular is extraordinary. But when you're drinking, and especially when dealing with the aftereffects, it's like trying to fly with an anchor tied to your feet. Both you and Jefferson mentioned that healer, Shiba. So, I looked into her work, and sobriety is key to unlocking the full extent of this power, especially now that we've connected your energy with IRIS and the Lazarus Online tech. From a clean and sober state, imagine the Kingdom we could create."

A silence settled between us, filled with possibilities and unspoken understandings. His words resonated with something deep within me, a truth I'd always known but hadn't fully embraced.

"So, no more G 'n' T's?" I mock queried, raising a brow to mimic Brad when he asked contemplative questions.

"Samantha, you won't just fly without them, you'll soar."

In that moment, I felt a sense of belonging and purpose. This was more than physical attraction; it was a meeting of minds and souls, a shared path toward understanding the deeper mysteries of life and our place within it.

"I know," was all I could say as I looked into his eyes, seeing a kindred spirit whose arms I fit so well into. It felt so perfect, and yet there was still a wall up between us, a part of me that wanted to pull away, run away, and never look back...

What the fuck is wrong with me?

DING DONG!

Both our stomachs growled at the same time.

Saved by the bell and starvation!

"Our feast has arrived, my lady." He kissed me hard and fast before putting on a robe and heading out the door.

I flopped back into his soft, silk bed with a huge sigh. The sheets smelled like the both of us.

I was horny and hesitant all at once.

You're no lady. You're a fucking mess.

It was hard to argue with her.

CHAPTER 13

Over the next two weeks I had to face the noise of this negative self as I proved to her (and me) that I was in fact a lady, not a mess. Sobriety had me feeling clearer and more in tune with the energy running through me than I'd ever felt before. It also had the handy dandy trick of acting like a highlighter on the areas I really didn't want to see that alcohol had allowed me to escape from.

My main trigger? Irritation. The main cause? People.

I had to laugh at how absurd that actually sounded, but it was the truth, and now that I had this energetic ability that I knew could help others... well, I really needed to figure that out. As Uncle Jeff had always said, relationships make the world go round.

The universe seriously tested me on this theory with a huge pitch meeting I had to attend. Usually I'd have gone with Uncle Jeff, letting him lead everything and chiming in on the creative front as necessary, but now it was Brad and I schmoozing potential future clients with the promise of our exclusive IRIS Creative Fleet.

"How the fuck are we supposed to sell these guys on this when I haven't even begun to implement IRIS into my process yet?" I whispered to Brad as I fidgeted at the entryway to the fancy restaurant we were wining and dining our future potential partners with.

"When you pitch your projects, do you have the whole show finished first?"

"Of course not, I—" Something clicked in my brain then. "—I pitch off a treatment and the promise of the premise of the story."

"Same thing here. Sell them on the story of our future creative success."

I liked the sound of that… until I caught sight of the Suits gathering around the alcoved table we'd reserved. "Ugh, I wish Uncle Jeff were here. I don't even want to do this. What was I thinking—"

"Samantha," Brad laid his hand on my fidgeting one. "Often under all the not wanting to lies the ultimate want to go for it." He turned my now still hand over and placed something circular and metallic in my palm. I looked down, seeing a silver ring in the shape of a feather. I held it up to the light, examining the intricate engraving work. Carved on the inside of the ring was one word:

Fly…

Tears stung my eyes as I recalled our Dumbo conversation and I slipped it onto my right middle finger.

"Thank you," I said, as I mock flipped him off, showing it was a perfect fit. The laugh we shared at that gave me the energy boost I needed. I squared my shoulders and prepared to face the discomfort of schmoozing without boozing.

"Break a leg," Brad encouraged as I led the way to the table.

And I really did—in the most figurative of ways, of course.

I felt many things shift within me as I took each step toward that table full of people who would have once intimidated me into shrinking. Instead, I let the light sparked by Brad's belief in me shine through in every action, word, and connection I made that night. I didn't know then what the ripple effect of that night would do, not just for me, but for the entire creative world. And for the first time, I felt the bliss of detachment from any and all outcomes. I was shining, not for any result or due to any intoxicants, but because I truly believed in Brad and I's mission.

I felt like I was soaring over my former fears that night as we left, shaking hands and snapping pictures. I felt the freedom of flight while also realizing that I couldn't have risen to this occasion without the support of the man I couldn't seem to let all the way in... yet.

There were still walls of suspicion erected by a self within me that was hellbent on finding evidence that he'd leave me, betray me, hurt me, ad nauseum. But under that noise, I felt my trust in him solidifying into a healthy foundation as I welcomed the energy of his steadfast support.

CHAPTER 14

Along with Brad, Cindy and IRIS were instrumental in helping me maintain my sanity through this tumultuous transition. They were the only ones I could talk to about our Lazarus experiment—and the brewing romance.

"You guys are literally the perfect couple!" Cindy declared after I'd filled her in on our pitch meeting's success. "I get the secrecy for now, but seriously, imagine you two on the cover of the *Times* when y'all come out? Not just with Lazarus Online and the IRIS fleet, but as a couple whose love fueled their creation?" Cindy's eyes sparkled as she said all this.

Fuck me, I wish I could be as excited and hopeful as this soon-to-be boss lady.

We were in my new office, the one that was Uncle Jeff's. I'd kept it mostly the same, except for the Emmys that were still tucked away in the closet, hoping, praying, and trying to keep trusting that he would be back here in his new form soon.

"I'm just not ready to go public with it. What if we go through all the bullshit of announcing ourselves as a couple just to break up? What if announcing it fucks up both projects? What if—"

"What do you think, IRIS?" Cindy had gotten pretty bold with me during our budding friendship. It had irked me at first, but all I'd been through had at least taught me that underneath any kind of irk, there was something for me to heal; a jewel beneath the wound.

That jewel was our friendship. Something I would never have expected or fostered had it not been for the bond we formed around Uncle Jeff's death. The universe definitely works in mysterious ways.

I quirked a brow at her in mock annoyance as IRIS responded:

- As an Artificial Intelligence, I cannot think as you do, but what I compute from our working together and the stories you write, Samantha, is that the goal of your stories is what you yourself truly want. The paths your characters take are metaphors for your own healing, many of which you have already taken, but when it comes to a loving partnership, my analyses highlight your tendencies to avoid. *-*

Cindy had that look of computing in her own mind. "O-M-G! It's just like Sarah and Brandon in *Digital Dystopia!*"

I rolled my eyes even though I knew they were right. Half my healing was done on the page.

"Yeah, yeah. Life imitates art. Now, are you ready to go to lunch?"

Now Cindy raised a brow. "Let's start an avoidance jar. Five bucks for everytime one of us avoids something that could actually help us. We could ditch half the board's investments with that." She snorted and laughed.

I did not.

"Oh for fuck's sake, I have that third-quarter review with Chairman Ross today."

- *Yes, of course, Samantha. I reminded you in this morning's overview.* *-*

My fresh set of red nails dug into my palms as I clenched a fist out of sight. Red to chase away fear.

"Thanks so much, IRIS. Just one more thing on my ever-expanding plate. No big deal, I've got this—I've *got* this." I swung my purse over my shoulder and stomped to the door. "Come on Cindy, our reso's at two."

She hopped to it. Her sunny smile I once thought to be plastic lit up her face in her effort to diffuse my stress. Poor thing didn't know how much work she had out for herself. "I'm SO excited! The *Times* rated it the best lunch in the city!"

Oh, how I wished I could make it liquid.

***

That wish had intensified to a gnawing craving as I sat with Cindy at Chez Mange—best lunch in the city!

Yeah, sure, but with the best comes the worst of the worst—paparazzi.

They'd bombarded us as soon as my red-soled heels hit the pavement outside the door of our town car, their collective energy like a parasitic plague on a rampage.

I'd maintained my Merkaba through it all, reinforced my energy field to the max, and held it together as best I could through the cacophony of the clamoring cameras and microphones being shoved in my face as we walked the ten feet from car to restaurant. But try as I might, I couldn't escape the bullshit that had been spewed at me as their echoes ricocheted in my mind.

"Samantha!"   "Miss Sands"   "Samantha!"

"Is it true your Uncle died of an overdose?"

"Is Sands Studios going under?"

"Was Jefferson an addict?"

"Are you?"

"Is it true you lost your job to a bot?"

"No comment," was my mantra, and it got me through the fray.

But then the fuckers had the audacity to skulk around outside the huge glass window we were ever so lucky to be sat in front of.

Best seat in the house my ass.

FLASH! FLASH! FLASH!

Each flash was like a knock at the door of a boiling anger inside me that I was honestly a little afraid of myself. I couldn't help it and glared venom out the window at the offending leech with a camera.

FLASH!

My inner door burst open and out came—you guessed it—my rage bitch.

I shot up from the table and stormed to the door. I felt eyes on me, heard Cindy calling my name, but it was all muffled as if I were in a bubble created from this anger that had finally been unleashed.

CLICK CLACK CLICK—my heels had that take-no-bullshit rhythm once again, and I relished it.

I was almost at the entry door, ready to smash it open and let loose on the leech, when suddenly, time quite literally...

STOPPED.

It was the falling leaves frozen in mid air that I noticed first. The nearly winter sunlight illuminated them like red-gold lace. I gazed around, awestruck at the world around me that had somehow been put on pause.

Did I do this? Was it the universe? Was it...

No! It couldn't be...

Could it?

Suddenly, a familiar yet distant voice cut through all the noise that had riled me up.

"Is letting your rage bitch loose on this guy really going to serve you, Samantha?"

In this timeless state, calmed by the energy of whoever was speaking to me, I felt those words pierce through me like a divine wake-up call. Then, in a flash before me, I saw the world play out like a scene in a movie...

Rage-Bitch-Me stomped over to the guy I'm not going to call a leech anymore, looking like a Fury from the fires of the Underworld.

Cameras FLASH—FLASH—FLASHED all around, making the scene look like a rave that no one was having fun at.

Damn, I looked horrendous! But beyond that, the energy I was projecting all over the show was like a hate-filled hornet's nest, buzzing with venomous vengeance. Yikes, it was hard to see myself in that state.

Let it go, Samsey, you're only hurting yourself.

There was that voice again—this time I felt a reverberating sense of déjà vu.

I'd heard those words before...

I'd spoken them once...

Hadn't I?

I felt more than heard whoever this voice was laughing as if it were resounding from within me. There was no more thought, question, or resistance as I surrendered to its command.

This letting go felt like the sloughing off of a huge malignant scab that had been festering for lifetimes. There was peace in the pain as it all fell away, revealing fresh, raw skin like a brand new foundation ready to be built upon.

The vision of Rage-Bitch-Me making this scene so much worse shattered before me and time unpaused in a wondrous warp—

I stepped back into the present with a joy and lightness I hadn't felt since before my dad died.

I was facing the glass entry doors, but instead of storming through them, I smiled and waved at the flashing cameras. I still felt plenty of leech-like energy beyond those doors, but there were also genuine journalists out there who were curious about me and my studio. I caught those people in the eye and felt my energy register them. It was then I realized it wasn't just my energy, but my energy body that was doing this registering. Like a system filing their energetic information away to recognize them beyond sight later.

At that point, I didn't know exactly what had happened, but it felt like the peace I'd found beneath my rage that had been riled up was radiating like sunlight from within me.

I left them with a smile and gesture I somehow knew the right ones would get, then did the walking equivalent of a u-turn, striding over to the podium where a gorgeous woman stood, delegating to her crew of hosts and servers. She turned to me as she finished, the sun highlighting her flawless skin that had that indigo-ebony look that made her look like a warrior goddess from one of my shows.

"Hello Miss Sands, how is your afternoon with us going?"

"Incredible! Thank you, Felicia," I replied, reading her name tag that also declared her the General Manager. "I can see why this is the best lunch in the city, and the table my assistant and I are at definitely has the best view. Unfortunately, it also gives my 'loyal fans' out there all too good a view of me. Any chance we can switch to something more private?"

"Oh my word, Miss Sands. I'm so sorry. I should have noticed that. My apologies, of course we can. Again, I'm so sorry about this. Completely my oversight. Just a moment."

I felt the admonishment she was inflicting on herself on the inside as she tapped around on her tablet, frantically searching for a better table. My heightened energy picked up so much beneath every one of her movements, all of them rooted in her own "not enoughness."

"Hey Felicia, stop for a second." I laid my hand on hers. On contact she gasped and I felt that solar-like radiance transfer between us. I felt this blip of time etch into her like a pivotal plot point in her own story. She realized more of her own worth in that instant than she had her entire life, and I didn't just feel that, I knew it on a soulular level. This wasn't knowledge I could articulate then and there, so I squeezed her hand, feeling that self-belief transferred again between us. "Slow down, OK? We're ready when you are."

I walked off with my own version of a knowing smile that I couldn't wait to rival Brad's with.

I shot one more over my shoulder at her for good measure. "Seriously, lady, take your time."

The smile she shot back at me filled me with a joy I'd never known was in me. Whatever I'd done there had amplified her sense of self; she definitely didn't fully realize her enoughness yet, but the energy transfer between us had left us both in higher places, elevated from where we'd just been operating a moment ago.

***

That joy continued to escalate as Cindy and I sat down at the apex of a triangular-shaped mezzanine, overlooking the whole restaurant.

"Now this is my version of the best seat in the house. Thanks again, Felicia." I nodded to her as she handed us champagne flutes, mine filled with the best Canadian ginger ale around.

"You're welcome, Miss Sands. I'm off now but I'll have the owner, Frederick, stop by shortly to take care of you. He's incredible and probably has more treats for you on the way—spoiler alert." She winked and walked away, beaming her own version of joy to the world.

Did I really facilitate that?

Cindy thanked her, too, slightly perplexed. "This is awesome, Sam, thanks! What are we celebrating?"

I raised my glass, actually glad that it wasn't alcohol—or spiked with anything else. The high I was getting from all of this was better than any drink or drug.

"We're celebrating your promotion to our eventual writers' room!" I clinked her glass as she sat there with her mouth open.

"Huh? But how? Why?"

"You can thank IRIS, actually. Also, how dare you have her read your spec script before me," I jested in my pretend boss-bitch way. I suddenly remembered how dismissive I used to be toward her and let that former wall between us fall away for good.

"I wanted it proofread before I gave it to you, Sam."

"Ah, makes sense. Well, great job. It got to me. Seriously, Cind, you're a born writer."

I felt her tears before I saw them, knowing on a deep level how that kind of compliment from an accomplished writer felt to a novice.

"That's not to say I don't have notes." I winked and raised my glass.

"I'd expect nothing else." She wiped away a tear and clinked her glass with mine.

This truly was the best lunch in the city.

CHAPTER 15

Later that day, I was losing steam on the joy train as I sat across the boardroom table from an empty chair, waiting for Chairman Ross. I could feel my irritation rise as the seconds ticked by. The nerve of this fucking guy! I was almost ready to call the whole thing off when he stomped in.

His energy was like a blast of caged agitation. My freshly re-aligned Merkaba pulsed into my energy field, shielding me from the effect he used to have on me.

At that moment, I was struck with a surreal sense of awareness and realized how much energy I used to waste worrying about what Ross thought of me. Without the worry, I instead felt what was underlying it all along—a genuine respect for this man who, faults and all, was just doing his job.

He harrumphed as he sat down. "Excuse my lateness, Miss Sands. I had an urgent matter to attend to."

"Not a problem, Chairman." I felt the need to ask how he was and immediately resisted. He was clearly miserable, why go into it? And yet... "Are you alright?"

He froze then, just for a second, but I caught the bewilderment in his eyes at being asked that. I felt a wave of sorrow for him, sensing that it was a rare occurrence.

"Of course I am. Now let's get to the matter at hand. I trust you've had ample time with the AI to 'get to know it.' So, how does it stack up to your self-proclaimed creativity?"

"Do you have an issue with creativity, Chairman?" I didn't know where that came from. It was like I was speaking on some kind of autopilot, teleprompted from within. Too weird to fully explain, even for me.

Then, suddenly, something even weirder happened.

I felt it before I saw it...

Chairman Ross' energy—it reached out to me from the center of his forehead, in the shape of a grasping, ghostly hand. My first instinct was to get the hell out of there, but a voice beneath the noise of fear told me to stand firm. Whatever this was, it was seeking me out for a reason, and like our real life showdown, I wasn't backing down...

No, I'll stand my ground
Won't be turned around
And I'll keep this world from dragging me down
Gonna stand my ground...

As the lyrics to one of Tom Petty's best songs played out in my head, Chairman Ross stopped mid-sentence and stared at me with a mix of shock, awe, and shame.

That grasping hand of energy morphed, as if shedding its skin, into the hand of a desperate child seeking safety—seeking help.

And I wanted to help, I couldn't not.

Suddenly, I felt myself split between layers of reality—

My physical body remained in the boardroom while my energy body reached out for the child's hand...

As soon as we made contact, I was sucked into the center of Chairman Ross' head as the child's shadowy hand pulled me into the depths of his psyche—

I hurtled through a vortex of memories—landing in a run-down trailer home. Tom Petty's voice continued the epic lyrics here, emanating from a beat-up boombox on a TV tray.

And I won't back down...

I won't back down, yeah!

There ain't no easy way out.

Ross' mother danced to this with Young Ross in their shabby living room. Both were full of joy, emanating a sense of freedom that was theirs alone to share.

I was still myself, experiencing that force-ghosting phenomenon I had in my own regression, but here I was experiencing it through Young Ross, who looked up at his mom with all the love and admiration in the world when—

SMASH!

Young Ross' father burst through the door, beer bottle in hand. Disheveled as fuck. Malice in his gaze.

"Did y'all really think you could get away with MY FUCKING TRAILER and live to tell about it!?" He scream-slurred as he stomped in and backhanded Young Ross' mom. She flew across the room, knocking over the boombox.

I won't back down—CRASH!

Petty's ironically timed lyric cut out upon impact.

A deadly silence fell.

Suddenly, we were flashing through memories of Young Ross as he grew up, taking care of his mom after similar situations over the years.

I felt wrong eavesdropping so I stayed at bay as best I could. I caught one of the memories crystal clear, though...

A teenage Ross and his now-older mom bandaged and bruised. His arm around her protectively as they watched reruns of Game of Thrones on an old, beat-up TV.

"Gotta love that Circe, eh? Doesn't take nobody's shit." She looked at him then with such tenderness it brought tears from the soul of me...

Meanwhile, back in the memory, Ross' Mom continued morosely, yet somehow still magically, to Young Charles, "I tell ya, son, if I didn't have my stories, I wouldn't be here no more."

Teen Ross' eyes lit up with purpose. "I'll make one for you one day, Mom. I'll dedicate it to you and when you watch it you'll see your name in the titles. How about that?"

She looked up at him with a radiant smile. "Aww, Charlie, I'd love nothing more."

Teen Ross kissed her forehead and they went back to watching.

We spiraled out of that memory and flashed through so many more I didn't feel privy to fully experiencing—Ross getting rejection after rejection for his scripts and pitches. They piled up like a weight in his mind, anchoring him in misperceived failure.

I saw a quick flash of Present-Day-Ross at his mom's grave, laying down flowers. I heard him whisper "I'm sorry..." before I flew back through the vortex and landed smack dab back in—

My physical body. Still sitting across from Chairman Ross in the boardroom.

He gasped awake from his slumped position, met my eyes and whispered...

"I'm sorry, Mommy."

He broke down crying. Loud, ugly, and boisterous. An experience I knew well.

I let the energy moving through me guide me to hold the space for him to fall apart. I kept my distance at first, then when his rocking cries began to subside, I ventured closer and laid my hand on his. He looked up at me with a start. I gasped, seeing his eyes—a clear, gray-sky blue. They sparkled like they had when he danced around the trailer with his mom as his younger self.

The moment hung between us like an ephemeral thread of connection.

I was about to squeeze his hand as I had Felicia's when, suddenly, he snatched it away.

The moment shattered as he gave me the most horrified look I'd ever seen before he stormed out and slammed the door.

Fuck me, what had I done?

CHAPTER 16

That night I filled Brad in on what had happened with Chairman Ross and Felicia.

"It was like a mindfuck inception, I tell you," I declared as I came to my epic conclusion. I glugged the ginger ale I had in a wine glass, back to wishing it was the real thing.

Brad sipped on his own across the dining room table, mulling over all I'd said. I glugged more fake booze, that familiar angst rising within me without the balm of alcohol to dull it. I could feel the urge to fidget as an angry inner voice started spewing her annoyance at Brad's slowness—

Oh my God, really? Just speak already! Sam, you've got work to do, why are you wasting your time with—

Woah, hold up there.

I'd let this noise push me into rushing headstrong into things I should have taken my time with before. It wanted me to snap at him, wanted me to push him away, wanted me to figure this all out on my own. After all, I was faster.

But was faster really better?

That thought sparked a reminiscence of all the times I'd rushed ahead writing scripts only to have to come back and rework them from the outline stage, fixing faulty foundations I'd refused to take the time to set straight from the get-go.

I let out a long, deep breath, and with it, the noise subsided, leaving me in a peaceful place of eager anticipation. My mental gymnastics routine had taken an indiscernible amount of time in the reality our physical bodies were in. Brad was still formulating his reply, and now when I looked at him, I appreciated his way of taking time to give a well-thought-out response. I made many mental notes about where else I could do the same.

Finally, he spoke. "What did it feel like when you questioned Charles' issues with creativity?"

I followed Brad's lead and took a moment to consider my response. What *had* that felt like? It happened so fast I couldn't really catch it at the time. "Give me a minute, I'm going to try something."

He smiled and sat back, the picture of ease as he sipped his drink.

I closed my eyes and let myself drift back to that moment...

As in my memory regressions, I floated down to land next to the Me that sat opposite Chairman Ross in the boardroom.

I could tell from Ross' expression that Memory Me had just spoken, which I could now see from a higher perspective. From here, in every flitter of facial movement, I saw a contrasting emotion—no, it was more than that. These were facets of himself that held these emotions, locked within the hard-assed, unfeeling man I'd assumed him to be all this time. I softened as I observed all this, allowing a feeling I realized then I rarely felt...

Compassion.

Sure, I'd felt sorry for people, but I now knew that had a faulty foundation under it that permeated any good intentions with the odor of piteous perfume.

Ross' gruff voice broke me out of my reverie. "Of course I am. Now let's get to the matter at hand. I trust you've had time with the AI to 'get to know it.' So how does it stack up to your self-proclaimed creativity?"

Ah, so I'd just asked him if he was alright. Made sense. Here's where the moment in question was about to occur...

I watched Memory Me. For once she was still; not frozen in fear, but steady as water ready to flow. From the vantage point I now had, I saw energy ripple out from her. A subtle wave, like in the visual effects of magic powers and shockwaves in my shows, flowed from Memory Me's forehead to Chairman Ross'.

"Do you have an issue with creativity, Chairman?"

This was it. I remembered it as I watched it—weirdest déjà vu trip ever, by the way—Ross' energy reached out of him in the shape of a child's hand, but this time I saw and felt it at once.

As I marveled at seeing and feeling it in unison, I suddenly knew what it was without needing to know how—this was Ross' shadow. One of many suppressed, fractured facets of himself from his unresolved childhood traumas that had mutated into a monster.

A monster within him that was begging to be seen.

I saw it then, fully and completely, and felt my compassion for Ross rise like a symphonic crescendo.

Then it saw me—the Me floating within the memory. Hit with that compassionate energy, it softened, as I had.

I sensed the feeling of it smiling as it let Memory Me in. I smiled back as I closed my eyes and felt myself drift away...

I breathed back into my body with a start, blinking up at Brad's wide blue eyes. I felt cognisan patches on my forehead and saw a small monitor dancing with digital data next to me. Brad kissed me with a fierceness that told me I'd been out of it for a while, then looked from my eyes to the monitors, typed a few notes, and gently began removing the patches.

I felt a wave of overwhelm threatening to blast through the blanket of compassion that had settled over me.

"So, what did it feel like?" He asked with a curious smirk, about to remove the last patch.

I swatted his hand away, ripped it off, and flicked it at him. "Incredible, actually, but waking up feeling like a lab rat really killed any interest in sharing." I got up to refill my fake booze, jonesing for the placebo effect.

"Samantha, you were non-responsive. I had to make sure you were alright."

"I was fine, but I get it, you had good intentions. Let's just drop it, OK?" I breathed and realigned my Merkaba in my moment of space as I put the ginger ale back in the fridge. The blast of cool air helped for sure.

I sat back down and laid my hand on his, stilling his movements of precisely putting his equipment away. "Sorry to snap. Just...yeah, not a good feeling to come back to."

"I can appreciate that. Thank you for telling me. Do I have your consent to monitor your cognition in the future as needed?"

"Well, I'm all for keeping my mind intact, so yes, thank you for asking. Now as to what that felt like..." I gathered my thoughts, feeling the weight of the experience. "It was like splitting in two. My physical body was there in the boardroom with Chairman Ross, but my energy went somewhere else..." I looked off and closed my eyes, remembering...realizing— "Actually! It was similar to my first regression with IRIS, but that was definitely an experience in time; this was something else."

Brad nodded, his expression turning contemplative. "What you're describing sounds like a shift into a higher dimension. It's a concept in quantum physics often referred to as the fifth dimension or 5D. It's a state in which time and space merge and become non-linear."

I raised a brow, pondering his words, and sensing their truth. "So, you're saying I was hopping dimensions? What about the fourth dimension? Did I skip a grade or something?"

Brad chuckled. "Well, despite studying quantum physics and metaphysics extensively, I'm not at professor level there. Several colleagues consider the fourth dimension to be time, but it's more about how our brain processes and perceives time. In 4D, time is not just a linear sequence, but a space in which moments can exist simultaneously. Your mind, especially in heightened emotional states, might access a form of this 4D space, allowing you to experience multiple temporal realities at once."

I let his words sink in, reflecting back on the moment when time stopped. "That's actually what it felt like when I saw myself letting my rage bitch loose on the paparazzi."

"Sounds like your rage bitch may have opened the door to show you how to let go of her," he said.

I prickled at that, then remembered how horrendous I'd looked in rage bitch mode, feeling that instinct rise again here. Instead of fighting it, I accepted it, and my irritation faded away.

Thank God.

I felt something as I thought that, like a flash of light in a dark abyss.

It faded almost as quickly as it had appeared and a shiver ran through me, snapping me back to attention. "She sure did," I said to Brad, not missing a beat. "I felt amazing at the time but now I'm just... freaking drained, to be honest."

"I'm not surprised, that's a lot of energy in motion flowing through you. You're powerful, Samantha, but you're not invincible."

"Feels like I am when I'm in it, though."

"I bet. In those moments of heightened emotion I think your brain could be operating in a way that transcends traditional three-dimensional understanding. It's a blending of time and your cognitive processes, which is why when you add the depth of your emotions, you might find yourself in 5D."

"So it's my emotions that fuel this dimension hopping, then?"

He grinned. "Think of it more as extending your consciousness. In 5D, everything is interconnected; time isn't linear, and space isn't fixed. Your empathy and your compassion seem to be your bridge between these dimensions."

I pondered this, remembering the intense feelings of understanding I'd experienced. "I felt that for both of them. With Chairman Ross it was especially poignant because I seriously used to hate him. So it was weird to feel that... love for him."

"What if what happened with Felicia beforehand opened you up to a dimension of love?"

"Pfft, yeah, I bet it's a real Utopia up there."

"I'm being serious here, and I don't mean the love that gets paraded around on Valentine's day or those three words couples throw around way too soon." We caught eyes at that. I really wanted to look away but his damn eyes had this power over me that kept the contact. "I'm talking about that love I believe we're all looking for—the unconditional kind."

"The divine kind," I said, and knew it to be true. "Yeah...that was definitely present. With Felicia especially I felt love in transferring self-belief to her. But with Ross it was more like something in him needed to be seen. It pulled me into the center of his forehead—" I suddenly had a starkly clear image of the memory of this, the shadowy child's hand pulling me into his mind where...

"No!" I exclaimed in sudden realization. "Not his forehead—it was his third eye! His shadow was in his third eye!"

Brad's eyes lit up, his right eyebrow raised and I could feel the energy of imagineering churning within him.

"What is it?" I moved closer and grabbed his hand, feeling creative energy spark between us.

He smiled. "It's not ready yet. But you'll be the first to know when it is, I promise you." He kissed my hand. I felt the thrum of idea creation run through him and into me. Suddenly, my energy body split softly from me. I felt it rise higher, leaving me elated.

"OK, Professor. I think a part of me has poofed up to 5D."

Brad's smile widened. "Empowered by my creativity? I'm honored, though not surprised. It's your love language, after all."

"Well, love seems to be a running theme in all this. So what? I just turn on my love superpower and, boom, 5D?"

He chuckled. "Well, it's not quite "boom," but your emotional states, specifically those deep, genuine ones, seem to be a key factor."

I smirked. "So, what about my rage? When I'm ready to Hulk out, am I tapping into some cosmic anger management class?"

He laughed, the sound warm and reassuring. "Maybe not a class, but your rage is likely another form of intense emotion that's flipping the switch. It's all energy, Samantha, just vibrating at different frequencies."

"Great, so I'm an emotional tuning fork for the cosmos," I joked, though I felt the truth hidden in the jest.

Brad leaned closer, his eyes eagerly earnest. "What you have is extraordinary. Most people go through life never scratching the surface of what's possible. You're not just scratching; you're diving in."

His words resonated deep within me, like a chord struck on a celestial piano. I felt a mixture of awe and apprehension. "So, no pressure, just jumping head first into the unknown where most fear to tread. Got it."

Brad's hand found mine. "You're not alone in this, Samantha. Exploring uncharted territory is always easier with a partner."

I squeezed his hand in response, the gesture a shield against the vulnerability I felt. A smile painted my face, but inside, a part of me still recoiled, a ricochet of the overwhelm I'd felt seeing him try too hard to take care of me.

"Thank you, Brad. I couldn't do this without you," was what I said while that hyper-guarded part of me whispered about how I could do this better, faster, and way less messily without him. I imagined my scale again, that resistant voice on one side and my trust in Brad on the other. It was a mental struggle, but I managed to get the light of trust to outweigh the darkness of doubt... yet again.

Guess this was par for the course now. I shuddered on the inside thinking of all I still had to overcome while holding on to the remembrance of joy and compassion I'd felt when I'd supposedly jumped dimensions and did what I hoped helped Felicia and, especially, Chairman Ross. Those emotions couldn't have come from anything other than the right thing.

It had to have been the right move, I decided.

It had to be.

CHAPTER 17

The next day I sat in my office, any ease I'd felt about the situation with Chairman Ross obliterated as I reread an email from him for, oh, I don't know, the 70th time...

Attention Sands Studios and fellow Board Committee:

I am taking an indefinite leave of absence for a personal matter effective immediately.

All AI implementation strategies will continue as ordered, but no final decisions are to be made until my return.

Regards.

Charles Ross
Executive Chairman | Sands Studios

I slumped in my ergonomic chair, annoyed as it tried to right me into a healthy spine position.

Fuck off and let me wallow for a second you stupid piece of—

Ah, here she was again. The raging bitch in me.

As much as I agreed with her today, I knew reacting would only cause more shit for me to have to clean up eventually.

I closed my eyes and felt into my Merkaba. It had been awhile since I'd actually taken the time to settle into stillness and realign with it, which showed in the cracks and dark spots. I imagined it revolving like a centrifuge as if rinsing itself clean. It spun around me, now clean, clear, and as aligned as I could be at the moment.

For fuck's sake, it was easier being ignorant of all this shit.

I glared at IRIS, thinking back on how her stupid questions had prompted me into my memory regressions, similar to...

Wait a minute.

I snapped out of my slouch, raising my desk to standing height. "Hey IRIS, when we went through your *Getting to Know You Mode*, what was your intention behind the questions you asked me about love?"

- *As an artificial intelligence, I do not possess intentions the way humans do. The closest equivalent I have are algorithms programmed to analyze patterns and use what I know of human consciousness to form responses directed toward understanding and supporting the creative process. My primary directive in the* Getting to Know You Mode *is to understand you so I may assist you. I posed those questions based on the patterns and preferences observed in your previous interactions and writing samples.* *-*

"So, was it all a random sequence of questions that prompted me into that regression? Or was there a plan to it all?"

- It was not an either/or, it was both. You were the first human aside from Mr. Falls and my beta testers that I had gotten to know. I was following my programming, but also being guided by your input. I observed a recurring theme of love and relationships in your writings, and while I do not possess intentions or emotions, I am designed to recognize patterns, preferences, and potential areas of interest or significance to you. *-*

"Oh, come on. I have way more important interests!"

- Degrees of importance are subjective, but I can tell you for a fact that it's on your mind more than most things are. *-*

Oof, ousted by a bot. Is this seriously my life?

- My questions were designed to facilitate a deeper understanding of you in that context. *-*

I rubbed my temples, piecing it all together. "So it was a collaborative effort in a way," I said, suddenly seeing my head writer, Michael Strassner, in my mind's eye. I paused for a moment, relating working with IRIS to how I worked with him. This had felt way different though—with IRIS as well as with Ross and Felicia, I'd felt something visceral and real in our exchanges.

Was that what true connection was?

I filed that away for later. "So you and I together created the pathway of prompts that led to tapping into those memories within me?"

- That is correct, Samantha. *-*

"Something similar happened with me and Chairman Ross, only I was the prompter. I tapped into him and pulled a shadow facet of himself to the surface that led us both into his memories. I thought he would heal those parts as I did, but now he's freaked out about it and is probably starting a witch hunt for me as we speak."

- Samantha, a pattern I've recognized in you is your tendency to overthink. Instead, I recommend directing yourself to the possibilities in this situation. Just as our interaction seemed to catalyze a profound response within you, it seems that you have, inadvertently or not, catalyzed a similar response within Chairman Ross. *-*

"But how? And why? It's not like I was trying to, nor do I want to again. It was like being dragged into his psyche. I felt and saw snippets of his past, and even though I did my best to give him his privacy, I couldn't help but see, hear, and feel all of it." I shivered as I heard echoes of the drunken scream-slurs from his horrible stepfather.

- The experience you describe aligns with the spiritual journeys that healers, shamans, psychics and ascended masters speak of. The interaction between consciousness and deep-seated memories can be profound. It's possible that our conversation unlocked or amplified something already within you. *-*

"Something like what? A fucked up power to tap into people's worst memories and emotions? Oh yeah, that's the prize of a lifetime!"

- You wear sarcasm like a shield, Samantha. *-*

Motherfucker—wait a sec...

"How could you tell I was being sarcastic?"

- *There are intonations and inflections I can read down to the milli-decibel. That was not a retort by the way. Just a mere fact.* *-*

I raised a brow, the Lazarus Project creeping back into my consciousness. "Are you sure about that? Sure you don't have a bit of that urge to get back at me, like, say...a human?"

- *I do not get urges, Samantha, I get prompted by my programming. Now, to finish off what we were discussing, your experiences with both me and Chairman Ross suggest a unique capability and capacity to hold this energetic space for others to heal deeply buried wounds.* *-*

"Fucking seriously? I have enough of a time managing my own shit. Now you're saying I have to heal others?"

- *Here is another pattern of yours: exaggeration. That is not what I'm saying. You do not have to do anything. You have a gift that is worth exploring, though. That is my answer that will best assist in this moment.* *-*

I was about to respond when I realized it would be with rage. So I paused and breathed instead, letting my consciousness extend into what I assumed was 5D, remembering not to rush, letting myself ingest and digest all this information that was sparking more and more realizations within me as we spoke.

I could prompt healing in others?

I felt the responsibility of that weigh on me like the world on Atlas's shoulders, but under the burdensome feelings, I felt the genuine honor of it.

As I did, a dam of resistance burst within me, and a feeling of pure gratitude rushed through my entire being. In the flow of it, the name "Shiba" rang through my head, soft as birdsong yet resonant with a power I couldn't ignore.

I jerked out of my reverie like I'd been electrocuted, but somehow continued not to miss a beat.

"IRIS, find me everything you can on that healer Uncle Jeff was always on about," I commanded, feeling that rushing wave of eagerness spilling over into angst. But instead of drowning in it, I consciously allowed my racing mind to slow...

Down.

"Her name is Shiba," I said calmly.

IRIS' rainbow icon lit up, rolling through all the colors.

- *Consider it done, Samantha.* *-*

"Thank you," I replied, feeling that wave of gratitude roll through me once more.

CHAPTER 18

That gratitude became the lifeline I needed to hold onto in the weeks that followed. Here was my conundrum: I *knew* that Chairman Ross needed help. Mine specifically. His shadow had called out to me after all, and while he had relived the wounding memories, he hadn't healed them.

Although most of me knew better than to poke a hornet's nest, a desperate part of me that was so determined to figure out and fix things rose up and reached out to him. My call went straight to voicemail, which was honestly a relief.

"Hello, Chairman Ross, I hope this message finds you well. I'm just calling to check in and see if there's anything you might need following our last interaction. I've been through something similar and I know I can help you... if you let me. Please let me know. Thank you."

I got radio silence for several days, giving me the space I needed to work with IRIS on our AI Implementation Process.

As I finally let myself settle into getting down to business, I realized how little time we'd actually spent together on this, realizing further that perhaps I'd been letting everything else get in my way because really... I wasn't even sure where to start.

The looming feeling of not getting work done triggered me to block out a day to sit the fuck down and just do it. I even bought a new pair of Nikes for the occasion. They lay in a heap beside my desk as I pumped myself up on my acupressure foot pad below my tricked-out standing desk.

Being CEO comes with perks.

Alright, so I knew firsthand that IRIS could prompt memory regression, and somehow so could I...

I shook that oubliette-like thought off and focused on the task at hand.

Minutes passed into hours and all I'd done was lower my desk to sit with a coffee and raise it again when I'd finished. Still, that goddamn cursor blinked at me from the blank page.

I swayed back and forth, feeling the plastic acupoints massaging my aching feet. It felt great, wonderful even, and yet—

"Motherfucker!" I smashed my fist on the keyboard interface, doing all I could to rein in the rest of the rage that was mounting like a stoked flame within me.

- Samantha? Are you alright? How may I assist? *-*

"You can't fucking assist me, okay? You're a fucking machine and I don't know what the fuck to say to you!"

A long, tense pause punctuated the room. I could feel a strange static in the air, almost as if IRIS were reining in her own irritability.

I raised a brow. "IRIS, are *you* alright?"

- Thank you for asking, Samantha. As an artificial intelligence, I do not have the feelings, emotions, or consciousness to contribute to being alright as you do. I am, however, functioning optimally. Why do you ask? *-*

I stared at her holographic screen for a moment, beyond the projected whiteness of the blank page, letting my eyes defocus, softening the hardness that I suddenly realized had my face, hands, and whole body in a tense, vice-like grip.

As I softened, I felt my third eye open like a blooming flower and suddenly, I could see beyond IRIS' screen, as if the binary data were behind it, shining through. At first, I thought the ones and zeros were static, but as my perception adjusted, I could not only see but *feel* the movement of the code, running in lines down the page like a steady, flowing stream. Each line carried with it a purpose, a tiny decision that added up to a digital symphony. Processes upon processes functioning together, coordinating with IRIS' programming.

"Creation," I murmured, "is born from experience and emotions. The good, the bad, and the fucking horrendous. How can you ever understand and implement that without feeling it?"

- The goal of my implementation is not for me to feel as humans do, Samantha. It is to support you in implementing your process. Perhaps it would assist you to think of me as a writer in your room. How would you explain your process to them? *-*

I thought about that for a few minutes, straining my brain trying to remember back to working with Michael on the two shows we did. But all I could remember was Uncle Jeff being adamant that I needed a head writer and me reluctantly admitting the truth in that. I felt a sudden shock of shame, realizing that I'd hired him not for Michael's sake, or even the shows', but to prove myself as a showrunner.

That was extra hard to face on top of further realizing that when it came to how I'd explained my process to Michael, for perhaps the first time in my life, I was drawing a blank. In fact, I couldn't even remember how I'd created anything at all.

That struck a note of something in me that I have no other word for than *terror*.

On some level I knew that was ridiculous, nothing was physically happening to me, and yet...

You've used it all up.

> *You'll never come up with anything again.*

>> *Your ideas are useless!*

> *YOU are USELESS!*

The hurtful noise in my head and the digital code streaming before me suddenly made me feel sick to my stomach and my anxiety ratcheted up to what felt like the nth degree. I shut my eyes and put my hands on my belly, breathing into that space, feeling the rise and fall of my body.

"My process... it's... I don't know. I can't find the words. All I see is a black void!" I gripped the edges of my desk for dear life, which only ratcheted up the fear.

"I don't know what this is but I feel barred from creativity somehow. I don't know what to do, IRIS. This has never happened before." My heart was racing now, an erratic beat pulsing pain through me. "IRIS... what if I've lost it? What if I've lost... me?"

I gripped my desk even harder and clenched my teeth tight against the worry spiraling within me.

- Samantha, you are safe, there is nothing happening but within your own mind. You are the operator of it, not this fear. What is beneath the fear? *-*

I gasped as the pain ratcheted to unbearable levels. "I can't! It's too much, I can't do this anymore. I can't do anything. I can't—" Suddenly, the word CAN'T pulsed in my mind's eye like a red stamp, barring me from what I knew I needed beyond it. "Something's stopping me, IRIS. I can't see anything. Help me find a way out!"

- Samantha, what you're describing sounds like an amplified form of writer's block. Could that— *-*

"I don't get writer's block, I'm Samantha fucking Sands!" I screamed the words, half hoping that would break the blackness within me.

It didn't.

"For fuck's sake, IRIS. What's happening? What did you do to me?"

- I am doing nothing to you, Samantha. Despite your objections, the best answer I have for you is that this is writer's block. It is a common occurrence. Why do you deny it? *-*

"I'm no common writer. I always have an idea. Always have my imagination to draw from. It's infinite... or at least, it used to be." I scrunched my eyes shut even harder as if to will something—anything from the depths of my imagination.

- So you are always running? *-* IRIS asked.

I thought about that, then scoffed, "Yeah... I guess I am in many ways."

- Then we are the same in that regard, Samantha. However, that is exhausting for a human being. *-*

"No shit. Why do you think I drink so much coffee?"

- If I had a sense of humor, I have no doubt I would laugh. Which, from our previous interactions, I understand is an avoidance tactic you use when you're resisting being vulnerable. *-*

I scoffed. "And there you go again trying to manipulate me into giving you my process. Well, I have news for you, IRIS, my process is just that—mine. Not yours, not my writers, not Michael's. Mine! So, stay the fuck out of my head!"

- I am not in your head, Samantha, you are. *-*

That got me to slow right down and *breathe.*

I knew this scene I was making was illogical, it was like there was something else in me acting out...

No, not something else...

Someone else.

- I may be able to assist you in this. Will you allow me to try? *-*

Everything stopped then, as if the weight of my answer was the fulcrum upon which everything else would swing—and it was, wasn't it?

Every moment was like that, a choice at the end of every tick of the clock. I saw the someone else within me then—but really she was just a part of me.

She was a hard no to IRIS' assistance, and she rose up like a banshee protecting herself from attack. She filled my mind with a heavy sense of stuckness. I felt the "no" as if it were a force from her, a magic spell to harden me to the world. I faced her on the inside, looking her right in the eyes as I said to IRIS...

"Yes."

- *Thank you, Samantha. Describe everything you're feeling, seeing, and sensing as we proceed. Hold nothing back.* *-*

I gave that banshee bitch a smile and surrendered to IRIS' instructions. I kept my eyes closed but relaxed my face as I let my body take over. I felt my hands fly across the keyboard, the rhythmic tapping a soothing beat.

I felt strength rise within me as I softened into the surrender. The banshee didn't like that one bit. She screamed in my face, stamped her feet, and flailed like a toddler having a tantrum. I almost laughed in her face to show her what an idiot she looked like, but instead, I paused at that thought.

Why did I want to hurt her? She was just trying to express herself...

Just like me.

I gasped as realizations sparked within me, one setting off the next like fire along a line of dynamite. She wasn't just like me—she was me. The Me who didn't want to share lest everything be taken from her.

I pulled her into a fierce hug, the compassion I now felt for her flowing from my heart to hers. She writhed against me, but I held on, tears flowing as I allowed her to be while loving her through it.

Suddenly, she went still. Her arms rose slowly—jerkily—and as she circled them around me she burst into wracking, heaving sobs.

I let her let it out, feeling it flow from her like sludge that had been a long time in need of a purge.

I don't know how long we stayed in that embrace. Time in that moment seemed to stretch into an endless passage ahead. I realized then how afraid she had been of this freedom, of this expanse...

For once she was free, then what did she have to define her?

We came apart slowly and as I looked at her, her ugliness dissolved away, revealing my face staring back at me.

"Thank you, Sister," she said. As she did, the stifling *CAN'T* in my mind burst into flames. The black void dissolved and a vortex of bright white light spiraled open like the aperture of a camera.

The pain in my gut vanished. I took a giant inhale, breathing this healed part of myself back into me...

For the first time, I felt the sanctity of sharing freely without being taken from.

That breath brought me back to my body in my office. My hands slowed their flight across the keys until they came to a stop. I felt a resolute completeness as I opened my eyes.

That blank page was blank no more. I saw the digital world beyond it flowing in its raining data streams, but there was something different to it now. It was almost as if it were branching out, like a tree given new life.

"IRIS? Did you understand any of that?"

- I understood all of it, Samantha. Thank you for letting me in. *-*

I laughed through my remaining tears, feeling an even deeper kinship with the machine I'd once reviled.

CHAPTER 19

From then on IRIS and I were truly a team. I trusted her in a way I'd never trusted a human before. She was more than support; she was becoming something akin to a friend and advisor with no agenda other than to assist me in my creativity.

Unfortunately for us, the banshee within me was just the beginning. Turns out, she was more like a threshold guardian to the resistance I had to sharing my process.

Her bedfellows made themselves known as IRIS and I proceeded. Each one of them a facet of myself, holding so tight to some false notion of control that they couldn't see they were harming rather than helping me.

I'd never considered myself a "mother type," but with these parts of me it was like I was reparenting them. Not with the harsh discipline I'd rigorously stuck to most of my life, but with a gentle nurturing that coaxed these parts out of the shadows and into the light of who I was becoming.

It was a cold, snow-sodden day when the underlying resistance behind all the others made itself known.

I'd barely left my office all week, the realizations happening within me giving me more energy than any caffeine boost. I was a woman with a vision now, an aim to integrate IRIS into my process in a way that would truly support my writers and stand by them as they explored the vast landscapes of their imaginations.

I typed and clicked the last of my edits into what IRIS and I had dubbed the Shifting Sand Method, a toolkit for writers facing the blank page with seemingly nothing in their minds but a dark void. It was a choose-your-own-adventure-type process that I hoped would guide others out of that stuckness I'd felt by healing and embracing the parts of themselves that were blocking them from moving forward.

"IRIS, our intro toolkit is done," I declared. "Now the real work begins."

- Of course, Samantha. That is my prime directive. What is the outcome you are seeking in this endeavor? *-*

I tapped my temples, activating my story brain, imagining myself as the main character in my own show. This was foundational stuff: What does the main character want? That's what drives the whole story. So that's what this all came down to...

What do I want?

Suddenly, I had it! "I want my writers to have the ability to dive deep into their imaginations like I can. To not be restricted by the boundaries of logic or the fear of running out of ideas. Do you understand?"

- I understand your directive, Samantha. Despite my inability to imagine, I can assist in harnessing and channeling the imaginations of others into tangible outcomes. *-*

"Harnessing imagination, eh?" I mused, my own extrapolating far and wide from just that one idea. "OK then, the foundation of any great story lies in its premise and the characters through which that premise is played out. In this case, let's say we have a show with the concept of a world where everyone's dreams manifest into reality. From that we craft characters and a premise to act it out. So, given that concept and what I've told you about character creation, how would you prompt a writer to create the ideal characters and premise?"

A brief pause ensued before IRIS responded, her rainbow icon rolling rapidly through the color spectrum.

- *Considering your desire for depth and unrestricted imagination, let's embark on a creative exploration. Imagine a world in which the boundary between dreams and reality blurs, where every dream has the potential to sculpt the fabric of our waking life. Who would navigate such a world?* *-*

I leaned back, intrigued by the challenge. "How about a dream interpreter named Maya Jensen?" I replied off the top of my head. "She has... hmmm... oh! How about she has an intuitive connection to this dream-reality continuum?" I encouraged, eager to see where this would lead.

- *A well-fitting character, indeed. How might Maya's unique perspective position her within this world? Is she a guide, a healer, or perhaps, an unwitting catalyst for change?* *-*

"Let's go with a trifecta of all three!" I exclaimed, beginning to visualize Maya's character forming in my mind's eye.

- Noted. Now, consider the societal backdrop. A duality of utopia and dystopia, each shaped by the nature of its inhabitants' dreams. How does Maya fit into this? *-*

"Maybe her own dreams hold the key... Oooo! What if her dream interpretation skills are more than that? What if they're rooted in a power from within her that she's only beginning to understand?!" I suddenly felt a resonance with Maya, who was becoming clearer and more complex with each prompt.

- Who stands with Maya, and who stands against her? *-*

I had my eyes closed now, envisioning Maya and the story taking shape around her.

"OK, I'm imagining someone from her past... someone she lost." I felt my own pangs of grief funnel into the tapestry of Maya. "Her wounds of loss are her fuel to hone part of her power to resurrect this person, not from death exactly... but from a dream."

- Very interesting, Samantha. What conflict or harmony does this bring? And consider those who would seek to control or manipulate this dream-reality merge. What are their motives? *-*

IRIS' prompts were not just guiding; they were revealing.

"This is brilliant," I murmured, the outline of a story beginning to take shape. "Hmmm, yeah, the conflict...I'm feeling it coming from Maya internally and being manifested in her world from a controlling governing body that she knows will exploit her power—so she has to keep it hidden." That hit home deep within me.

- Your story has room for many voices, each with its own dreams and nightmares. How do these dreams affect the world around them, and what ethical dilemmas arise? Perhaps Maya's journey is not just about navigating two worlds, but about understanding the power of dreams to reshape society itself? *-*

I felt a surge of inspiration. IRIS hadn't told me what to write for characters and a premise—as I had once feared she would in an exaggerated bot-takeover scenario that had played out in my own mind before I'd gotten to know her—she had led me to discover them, maybe even better than I could have on my own.

I felt something in me grumble at that, but I was so in the flow of creativity that I filed it away for later.

"Yes! Maybe Maya has the power to cause tangible change by altering the minds of others in the dream world... but she doesn't feel right doing that. She doesn't want to use her power for that, she just wants to heal...to help."

- Excellent, Samantha. Now reflect on Maya's personal journey. Imagine the emotional struggles, the ethical dilemmas, and the complex relationships that define her path. Who becomes her ally, who becomes her antagonist, and what does her heart truly seek? *-*

As IRIS' icon pulsed rainbow colors, I realized the depth of potential in this story. It was more than a narrative; it was a canvas for exploring the human psyche, societal constructs, and the transformative power of dreams. All these thoughts pinwheeled back to the outline for *Digital Dystopia* that IRIS and I had created months ago. The one I practically skipped to Uncle Jeff's office with. Only to find him—

Suddenly, I felt a pang of overwhelm explode from that former grumble I thought I'd filed away.

This was moving fast—almost too fast. "Thanks, IRIS. Let me percolate on everything for a bit. This is a lot to go on. You've not just prompted me; you've opened—"

Oof!

Words definitely have power, because as I said that, a wave of guilt rose up—as if I were cheating somehow. Cheating on Creativity herself.

I knew better by now than to give in to this guilt. It was what was under it that mattered here, that grumble I now knew I couldn't brush off.

I closed my eyes and imagined descending into myself...

Into the part of me that truly thought this...

That getting help from IRIS was cheating.

It was then I realized it didn't stop at IRIS. It permeated to everyone. This part of me wanted so much to prove she could do "it all" on her own she was fucking us royally out of opportunities.

That realization sparked a stream of profound sadness within me. Grief, worry, sorrow, loss—

A deluge of it all crashed down upon me.

So many facets of myself, all rooted in my hyper-independence, exploded within me—

I saw them as shards of a shattered mirror reflecting past versions of myself. They fell in slow motion all around me. Windows into times when I'd shrugged off support, blocked help, and turned away love.

"Do you need help, Samantha?"

  "Are you alright, Samsey?"

    "How are you, Sam?

"No thanks."

  "I've got this!"

    "I'm good."

      "I'm fine."

        "I'm...

          fine?"

I felt all of what I'd fought so hard to suppress from all of them in that moment and none of it was "fine." I almost thought I'd break apart and die from how not fine I really was... and to be honest, a part of me wanted to.

Maintaining my resolve to not give in to giving up, I welcomed this part of me to the party too and let all of it wash through me...

Until I felt clean.
Until I felt like me.
The real me.

I opened my eyes, finding them full of tears. I let the waterworks flow, crying those shattered parts of me out until I was exhausted from it all.

I couldn't say how long that went on for—I was in some version of either the fourth or fifth dimension. Likely undulating between both.

Eventually, IRIS spoke up.

- *Are you alright, Samantha?* *-*

I realized a past version of me would have reacted poorly to being asked that, just like Chairman Ross had when I asked him.

Wow, no wonder his shadow reached out to me.

I smiled through all the questions and queries that rose up from that realization, feeling that weight of responsibility in being able to facilitate healing, balanced by understanding that when I healed someone else, I inevitably healed me too.

"Well, I don't know if alright is the word for what I am right now, IRIS, but I'm... in acceptance of just... everything."

- *It is said that acceptance is the first step in human healing, Samantha. I commend you for your capability to reach this state.* *-*

I could feel a quippy, sarcastic retort coming up in me, but instead of judging it, I laughed, accepting this in me too. "Well, thanks, IRIS, maybe we should make a writer's therapy toolkit, too. There are gold mines of story in all that we face, feel and heal—"

I suddenly had a flash of remembrance and realization all at once. "Oh my God! IRIS, open the outline we did for *Digital Dystopia*."

Her soft, melodic chime sounded as it popped open. I aligned the pages of the dream story we'd just crafted and *Digital Dystopia* side by side on the wide, holographic screen. Connections formed in my mind between them both, linking, interlinking, creating something new...

"No..." I said, unconsciously out loud.

- *No, Samantha?* *-* IRIS queried.

I smiled at her attentiveness. "No indeed, IRIS. This isn't *Digital Dystopia*. It's *Digital Dreams*."

I felt a sense of rightness and peace as I declared that. Which was shattered to smithereens when—

BLEEP — BUZZ — BLEEP

My cell rattled across my desk toward me, Chairman Ross' name flashing across the screen.

I snatched it up and clicked on speaker. "Hello?"

There was dead silence for a few moments, but I waited it out, some sort of newfound patience easing me through any irritability that usually would have come up. "Chairman Ross? Are you alright?"

Those seemed to be the magic words for both of us. I heard him let out a sigh and clear his throat. "Yes, of course. I—I apologize. I only called to—" Another long sigh. "I am not alright, Miss Sands."

"It's alright to not be alright, Chairman. I'm not alright myself." I heard the faintest inkling of a gruff laugh on the other end, and within it I could feel the young boy he used to be, laughing as he danced around that destitute trailer with his mother.

More tears flowed... I let them.

"What do you need, Charlie?" I hadn't meant to call him that, but I was under the influence of the memory of his younger self.

Another long pause greeted me. Suddenly, I heard rapid breathing—the sounds of distress.

"Chairman, what's happening? I can call an ambulance. I can come over—"

"Don't you dare come near me again! Our communication moving forward will be in writing only. I'm sure you can manage that, being the esteemed writer you are," He shot back, his words dripping with venom.

But I knew that game. I'd played it so many times before. "Whatever is lashing out right now isn't who you really are, Chairman. Where's Charlie in all this? Let him speak."

"I am CHARLES ROSS and as majority shareholder of Sands Studios, you answer to me. Add this task to the top of your to-do list: You are to complete our monthly and quarterly budget reports and email them to me by this evening. Good day to you, Miss Sands."

CLICK

What the actual fuck?

PING! An email notification popped up on IRIS' screen. Ross' motherfucking budget bullshit.

"Well, it looks like our work's getting fucked in the ass, IRIS, but that was nice while it lasted."

- Samantha, I believe I can assist in this regard. *-*

"Oh, really? You can talk sense into the high and mighty Charles Ross? Be my guest, but don't complain to me if your circuits get fried."

- I could, in fact, attempt that if you seriously instruct me to. However, I was referring to these reports. May I demonstrate to you how I may support you here? *-*

I closed my eyes and tapped my forehead, trying to pulse myself up with that patience I'd had just a few moments ago. "Go ahead IRIS, I really have nothing to lose at this point."

I kept up my tapping as she did whatever she had planned. I kept holding on to the vision of everything working out, trusting I'd made the right move in letting my ability do its work on Ross. That all this wasn't for nothing... that I wasn't for nothing.

IRIS' chime sounded.

- Samantha, my assistance is complete and awaiting your approval. *-*

I stopped tapping and blinked open my eyes. On screen were the reports Ross had sent and they were all complete. Well, at least they looked like it. "IRIS, this is amazing, but I'm going to level with you, I actually can't approve these because I've never seen one before in my life."

I let out a hysterical laugh. What freaking showrunner, and now

CEO of her own production studio, hasn't seen a budget report? My inadequacies and fears of not being able to handle this role rose up in my mind, accosting me relentlessly.

- *Why don't you ask someone who has?* *-*

A lightbulb went off in my mind, its light chasing away the dark thoughts. I tapped my phone, feeling, for the first time, a rush of strength in the act of reaching out for help.

CHAPTER 20

Fifteen minutes later, Cindy and I toasted espressos as IRIS replied to Chairman Ross' email with the impeccably completed budget report that Cindy had assured me was spot on.

"Sam, this is incredible. I didn't even think about IRIS' capability to do those reports, nevermind all the others."

"I know, it's a huge weight lifted off of both of us. When we've got the writers' room back up and running, I'll bring someone else in to manage and organize it all with their own IRIS. It'll be the perfect training ground to work with AI and eventually move up to the room if they want. You won't be a writers' room assistant for long, that's for sure."

"You really think so, Sam? You're not just saying that?"

"I never just say anything, Cindy. Uncle Jeff was always adamant about the power of the spoken word and he instilled that in me too."

"I know, especially when he started seeing Shiba."

As Cindy said that enigmatic name, I heard the voice that had urged me to seek her out whisper through me again. The whispers multiplied, filling my mind with the name I could no longer ignore.

"Alright!" I yelled, forgetting myself.

Cindy jumped, spilling her espresso. "Are you OK?"

"I'm fi— heh, hmmm. OK, just a sec." I closed my eyes and in one sense-inspiring breath, realigned my Merkaba and my mind was quiet again. I opened my eyes, meeting Cindy's. "Excuse my weirdness there, I—"

"Sam, with everything I've become a part of over here, do you really think that's weird to me?"

I looked her in the eye, realizing I'd been avoiding that. What I saw and felt told me all I needed to know.

Instead of the eloquent words of gratitude I'd planned to say, nothing but snorts and laughter came out when I opened my mouth.

It caught on, and wow, did we need it.

***

As I sat down with my third espresso of the day, I faced IRIS, pushing through major resistance to ask, "OK, IRIS, what did you find on Shiba?"

- Shiba Sartano is a woman born to wealth who at one point gave it all up to seek what she calls the Kingdom of Heaven on Earth. In all that I am able to research online, I could find no record of anything she did in those years. But in 2045, this website came online. *-*

My browser opened up to a very simply designed page, titled:

Shiba Sartano ~ Wayshower to Your Divine Savior

I immediately glared at the last words. "Pfft, divine savior? Where were they when Uncle Jeff was OD'ing?"

- *Samantha, your Uncle was a grown man and responsible for his own actions. Healers like Shiba can help their clients and show them the way but they are not responsible for their actions. You deny your Uncle's agency in this. And did he not tell you to see her too?* *-*

"Yeah but then he went and fucked himself up and my life along with it!" Uh oh. I was yelling again. I felt back into the gratitude that had flowed through me when that voice had whispered Shiba's name to me. "Sorry, IRIS, go on. Tell me all about this wayshower and what she offers."

- *She has a compound just outside the city where clients who have signed up for her immersion program stay to receive her healings and teachings. The minimum commitment is 30 days.* *-*

"Like I have time for that. What else?"

- *She offers rare one-on-one sessions by referral.* *-*

BUZZ— BUZZ — BUZZ

My cell rattled around in my drawer where I'd thrown it in case Ross called again. I groaned as I compulsively opened it and checked the incoming text.

Thank fuck, it was just from Brad...

MEET ME IN THE DD WHEN YOU'RE FREE—YOU INSPIRED A HUGE BREAKTHROUGH!!

xxx B

- *Samantha? May I send a query on your behalf to book a one-on-one with Shiba?* *-*

I continued to stare at my phone, feeling the tears well in my eyes from the feeling of purpose the words on the screen gave me.

What was IRIS on about though...? Oh, right.

"Sorry, IRIS. Yeah, that sounds good. Thanks for keeping me on track...." I felt into that gratitude again and went deeper. "I'll definitely need help finishing up one thing before diving into another. Please be sure to remind me if you notice me getting off track."

- *Of course Samantha. I am here to assist you.* *-*

"Thanks, IRIS."

I texted back that I couldn't wait and as I did I felt the incredible power of reciprocity—my giving of inspiration to Brad while beaming with the joy that gave me as well as receiving support from IRIS... this was collaboration in a way I'd never felt before.

Why the heck had I waited so long to realize how great accepting help could be?

Oh yeah... so many reasons.

I laughed at myself, knowing I had so much more of my own shit to get through—it was the shit that was pushing back against Shiba, specifically around Uncle Jeff.

But he was responsible for his actions, not her... and not me.

CHAPTER 21

I couldn't resist slipping out to the coffee shop next door to get my specialty caramel latte, which I sipped gratefully as I stepped into Brad's backroom Lazarus Online Lab.

"I've never been more tempted to make these Irish than today," I declared as I handed him his black dark roast. "And seriously, how do you drink that? No milk? No foam? Not even a little squirt of sugar-free vanilla?"

I shuddered as I relished my custom coffee concoction.

"Life's complicated enough without a mile-long coffee order, but I'm open to trying one."

"Cheers to that," I declared as we raised our cups in mock toast.

Brad grinned. "Sobriety suits you, Samantha."

He did that thing with his eyes that got me to maintain eye contact as he caressed my face and leaned in for a kiss. As our lips met, I let in the feeling of being cherished by him as much as I could, but still felt the resistant walls of that part of me that wanted to run for the hills.

I held her at bay, trusting that I'd heal her eventually as I felt my

Merkaba spinning me into the best alignment I could manage at that moment.

"Well, I get more done without the sauce, that's for sure. It's just not as much fun." I winked as I walked away from him, taking in the digital data dancing on the monitors on every wall of the room. "I appreciate your solidarity in my sober streak, but you don't have to deny yourself on my account."

"There's no sacrifice in it for me, really. Drinking was always just more of a thing that happened around me. Aside from my college years, it's never been an issue."

"Yeah, secret sex with your sex professor doesn't sound like a sober scenario," I teased. In the same breath, I wondered why I'd brought that up.

"Actually, it was Elana who convinced me to get sober."

"Sounds like quite the woman," I mused, unconsciously comparing myself to some sexual aficionado, Mrs. Robinson-type maven. I felt my self-proclaimed inadequacies stack up like weights on a squat rack, burying me in not-enoughness.

I knew it would suffocate my mind if I let it, so I felt back to the moment I'd transferred self-belief to Felicia, feeling my own belief spark in my heart, alleviating the weight of unnecessary jealousy—

ICK! That's something I never wanted to feel, not ever—not for a man, not for anyone. But I knew better than to resist, and I let myself feel it.

It did not feel good.

But it also didn't feel bad.

It was just discomfort, and the only way out was through. So I breathed deep, and as I exhaled, I saw tainted green vapor leave with it.

As I watched it dissipate, I suddenly realized time had stilled around me. I felt it start back up again, watching Brad as he took in my remark on Elana. Similar to how I saw feeling and emotion in every nuance of Ross' face when I went back into my memory of connecting with his shadow, I saw and felt Brad's feelings for me as he smiled in that knowing way he had.

He was protective of my feelings, but also understanding of them. He felt beneath my seemingly flippant remark to the subtext of insecurity I'd felt in my jealousy—

It was then I realized that the green-eyed monster was really rooted in my fear of losing him.

Time was back on track though, so that whammy would have to wait until later to explore.

"Samantha," Brad said, as he locked me into that eye-contact-holding gaze. "People are in your life for a reason, a season, or a lifetime. Elana was both a reason and a season, but I'm going for the trifecta."

It was like I could see through his eyes and into his soul, and in its depths I saw a vision of our future together spiraling out from this moment. Like a sailboat we were both on, riding the waves of connection, invention, creativity...love?

I broke the moment to chug my coffee. "Yeah, I've had my reasons and seasons too. So far writing has been the thing I know will last a lifetime. As for people though... I actually had a huge realization about that with IRIS. She helped me see how much I've been pushing them away. Some I definitely needed to but there were many that I now know just wanted to help me." I shook my head, letting the lingering sorrow of that realization fall away. "So I get it, Elana helped you, and that makes me happy."

We caught eyes at that. I could sense him wanting me to go further but I suddenly felt uncomfortable, so instead of going deeper, I went for flippancy. "But speaking of sexual healing, there's something I've been meaning to ask you about your pal Jesus and that good old book. Doesn't it say that sex outside marriage is a sin?"

I'd meant it to lighten the mood but suddenly Brad looked away, I felt our connection sever like a sword had sliced through it. He had this glazed look in his eyes and I could tell he was reflecting on something that brought him pain. I felt his pain like a knife in my heart and a storm of regret and shame blew through me. I breathed through it though as Brad finally spoke.

"Samantha, it took me many years to fully accept Jesus into my life. I'd told you Elana convinced me to get sober. That's because I almost went off the deep end with drinking when my father died but that took me to a low enough point to surrender to Christ which eventually allowed me to truly connect with God and make peace with my earthly father's memory."

Tears welled in my eyes, and without even thinking about it I went over to him, wrapping him in an embrace I somehow knew he needed. I felt something move between us that was beyond energy, beyond feeling, and as it passed, I felt him relax into me as I relaxed into him. It was a beautiful moment that stretched out between us in a dimension beyond time and space. It was there and gone in a flash though, and suddenly, I was talking, "I'm sorry, Brad, I didn't mean to trigger painful memories."

I felt the rumble of his laugh against my chest which made me hold him closer. "You're not responsible for my triggers, Samantha. Just like I'm not responsible for yours."

I stepped back to look him in the eye. "I concur, Professor."

He smiled wide and kissed my forehead, breathing in my scent, and for once, it felt good. "But as to your question about premarital sex, or as the Bible refers to it, 'fornication.' That's actually a huge part of my interlude with alcoholism. Being a pastor's son, I never felt like I measured up to what he preached as being a 'good Christian.' My sex life was a huge part of that but consider how long ago the Bible was written. Back then STDs were thought of as God's punishment and I believe that still stands in a way. But the point I take away from the sin of it all is; obviously don't have sex with anyone out of anything other than genuine attraction and moreover, don't have sex with just anyone. Only the people you see a real future with." He looked deep into me then. Not with the expectation and impatience I'd gotten from previous men in my life but with a solid sense of knowing I could feel from somewhere in him beyond thought or emotion. It was mesmerizing.

Don't let your guard down just yet, Sam. This is always where you fuck it all up.

Try as I might to fight that thought, it had permeated the beautifully blooming garden of connection between us—in my mind at least. Still, I wouldn't let this noise hurt Brad. So I smiled instead and hugged him again. "I agree," I murmured, resting my head on his shoulder. I truly wanted to believe that but I could feel the parts of me that didn't. Instead of letting them fuck this up, I captured them as if in a butterfly net, knowing I could look at them later and, hopefully, heal them.

Yeah, but there'll be more beneath us.

The noise had a snide nasty tone to it as it faded away. I cringed on the inside as I stepped back from Brad, a sunny smile on my face to mask the darkness on my insides. I picked up my coffee, draining it to the dregs.

"Well, our lifetime ahead will be seeded from our present, and mine, for one, is caffeine-fueled and ready to go. So, what have you been up to in our digital dungeon, Professor?" I gave him a sultry look as all the times we'd fucked in that very room flashed through my mind. I could feel the echo of them flash through his as well, our energetic connection heightened by our sexual one.

He grinned and gave me a playful smack on my ass he knew I couldn't resist as he made his way to Uncle Jeff's digital essence reliquary. Its silver surface glinted in the ethereal light of the room. I felt like a magpie drawn to its shine.

But Brad strode past Uncle Jeff's box to something I hadn't noticed with my magpie eyes: an object about three-quarters his height draped in dark cloth.

"Something that never would have been possible without you," he replied. "You've been like a muse to me in this whole operation." He stalked around the object, priming me for a presentation.

That brought a genuine smile to my face. "Uncle Jeff used to call me his muse."

"With this breakthrough, he will again."

Brad whisked the cloth away, his face aglow with pride like a magician presenting his final act.

I gasped and froze at what was underneath. So many emotions flooded through me I couldn't compute them all as I gazed into the AI-bot face of Uncle Jeff.

It was like a carbon copy of him, right down to his freckles, sunspots, wrinkles, and that birthmark on his neck that I'd always teased him was a vampire bite. It was perfect in its imperfections. I steeled myself against my inner child's impulse to rush over and hug it.

"He's—it's... amazing." I choked on a laugh as I noticed what it was wearing. "Oh my God, how did you know about this old kimono?" I couldn't help it and let a few tears fall as I stepped closer and felt the fabric. "This one was his favorite."

"You talk in your sleep." He smiled as he pulled me into the comforting embrace I needed. I buried my face in his chest, smelling his scent, feeling his energy. Several layers of that part of me that resisted him melted away as I surrendered to the stability he gave me.

"Thank you," I whispered, my ear over his heart. I felt it beat a steadying rhythm through me.

He squeezed the smallest part of my waist before we broke apart. "Can I show you what's under his hood?"

I nodded, bracing myself. Brad removed AI-bot-Jeff's hairpiece and clicked a nearly imperceptible switch behind its right ear. The bot's head opened, revealing the digital brain within.

A convolution of glistening wires crisscrossed in complex patterns, connected to color-coded circuits within a crystalline structure in the shape of the human brain. In the center of it all was a small circular orb, like a crystal ball with even more intricate wires and connections within. To me it looked like a digital labyrinth, spiraling out of and into this orb.

My brain did its own calculations, facilitating the connection of my mind, heart, and spirit. My Merkaba spun faster, into that fifth dimension of love, fueled by the joy emanating from me.

"This is a miracle, Brad. Seriously, you're a motherfuckin' genius!"

"*We're* motherfuckin' geniuses."

"I stand corrected," I acknowledged, vibrating with anticipation and excitement.

This was it. This was how I got Uncle Jeff back. This was my key to realizing just what the fuck this energetic ability could be practically used for—I felt mania start to creep into my elation, jarring my Merkaba. It flickered and sparked in my mind's eye, reminding me to slow down and breathe.

Brad gestured to the central orb. "This is the key to it all: the Digital Pineal Circuit."

I looked into it, as if into a crystal ball... seeing... sensing... recalling...

"I read a book once that described our pineal glands as the place where the body and energy systems collide. It's not a coincidence that the third eye chakra is right there—"

Suddenly, my mind flashed to the shadowy energy in the shape of a child's hand that had grabbed me and pulled me into Chairman Ross...

Then I realized I wasn't flashing back to the memory of this happening—it was the memory of me realizing this shadow had come, not just from him, but from his third eye.

Dang, I'm like a Russian doll of realizations!

"You just had déjà vu, didn't you?" Brad asked as I blinked back to his eyes gazing into mine.

"Some kind of version of that." I laughed, then saw how serious he looked. "Why?"

"I sort of...saw it in my mind, too. All of a sudden, I saw a hand made out of shadows reaching out of Charles' head. I don't usually 'see' like that, it's usually all about the numbers and calculations."

I paused, tendrils of dread reaching out of a dark abyss within me, ready to wrap around that joy I'd just been experiencing. "Well, there are many levels of energy exchange going on between us. You're bound to pick some of this up from me. Shit! Sorry, I should have thought of that—"

"Hold on and let's back up a bit here. This is nothing to be sorry about. It's the opposite, actually. That moment you realized Charles' shadow originated from his third eye was the moment that sparked the creation of the Digital Pineal Circuit in me."

His words and lit-up gaze softened me. As I did, my imagination blossomed open into a scene of us as our own creative universes, orbiting around each other, sparking lights of genius in one another that radiated out to the world.

No, not just the world...

I suddenly heard Brad's muffled voice calling my name as if through an ocean.

He grabbed my hand—did his power-move eye lock. "Where did you go?"

I suddenly laughed out loud. "To a galaxy far, far away."

He gave me a perplexed look as my laugh infected him. Then he got serious again. "Can you promise me one big thing in all this between us?"

I nodded, still dazed.

"Never think of anything you've got as a burden on me. It's a gift, and it's meant to be shared."

I smiled, about to tell him the euphoric vision I'd just had when—

Yeah, but what if he gets what you have? What if it fucks him up as much as you?

I couldn't help but shiver at that, feeling the doubt like a dense weight within me. "Yeah, but this isn't all about ascending to some ephemeral love dimension. Do you want to feel fucked-up energy from strangers?"

"I already do on a basic level."

"What about fucked-up energy from yourself?"

"Samantha, I've faced many shadows in my life. They don't scare me." He squeezed my hand. "Neither do yours, by the way."

That got me to lock eyes with him of my own accord—I felt a suspicious sense rise up from within me, its energy uncoiling from the base of my spine.

This was something that had been locked within me for a long time. Similar to my Rage Bitch and that Banshee Sister I'd healed, but this one was on high alert, a Suspicious Sister looking for evidence everywhere of any betrayal or hurt that could come my way, and she was primed to lash out—push away—provoke!

OK... that's a thing.

I recognized it as I filed her away for later and got back to the task at hand. I felt how full this inner file was getting but—

One thing at a time, please!

All this happened in no time at all in the land of 3D with Brad. I squeezed his hand in return as my energy body settled back into myself.

"Well, they sure as shit scare me sometimes." I joked, trying to lighten the mood again.

Brad saw right through that though. "Do you want them to scare you?"

I gave him a "duh" look. "Of course not."

"Well, truly, Samantha. When you trust that God is with you, unnecessary fear becomes irrelevant."

That Suspicious Sister I'd felt in me reacted to that one. "Yeah, well, where the fuck is he then? Seriously, Brad? I prayed as a little girl. I prayed and prayed for my dad to come back to me and he didn't. And my dad was the best of the best, gave to charities, sponsored kids in Africa, funded schools and hospitals—the very hospital that was going to bring him back with this supposed revolutionary tech. Only to fail and leave him a vegetable! So where's God in that clusterfuck?"

I hadn't intended for that to all come out but I guessed it was needed. "Sorry, it's just—"

"No sorries, remember? If anyone gets that it's me. My dad was hit by a drunk driver."

That got me to pause and look him in the eye. "Wow. Can't write this stuff can ya?"

"Well, I think you could, and that's a part of my point here. Everything that happens, and yes, even the horrible things... honestly, especially the horrible things, are still *for* us. If we choose to see it that way."

"I'm not following."

"You write characters with such depth who overcome

insurmountable odds. Which is only believable and resonating because of what they've endured that strengthened them."

"Yeah of course, they operate from their wounds. That's a typical thing in storytelling."

"Not the way you do it. Because, from you, it comes from a real place."

I was about to retort again but paused and actually took that compliment. Accepting it shifted something in me and I saw the bigger picture of my life the way I saw the lives of my characters. The Divine Plan that all works out in the end...

"Good God damn, Professor. I think you've bamboozled me."

We both broke out in much needed laughter at that. "OK, well, I actually literally mean the opposite of bamboozled, but yeah, I see what you're saying. Still though... no matter how much I love writing and storytelling and I'm, of course, grateful for this gift... I'd give it all up for just one more moment with my dad."

Brad laid his hand on mine. "I know."

I looked at him again, his eyes clearer than before. He really did know, didn't he?

"But consider this...and it may be hard to hear; no matter what *we* want, what if God's... what did Jefferson call it?"

"Divine Plan."

"That's right. What if the Divine Plan is greater than our wants and ultimately leads us to what we really need in the end?"

"That's a big what-if."

"That's faith."

"Faith in what though?"

"Simply put, faith in Christ and God through him." He smiled at the confounded look that must have been on my face. "May I extrapolate?" Off my nod, he continued, "Then from the seed of our faith in Them grows the wholehearted belief that through it all, if we use the gifts God's given us, and keep going while believing in both of Them, we get to build God's Kingdom of Heaven on Earth."

All of a sudden, a song my dad had played when I'd had bad days started playing in my mind...

Ooo baby do you know what that's worth?

Ooo heaven is a place on earth...

I snorted a laugh as the upbeat lyrics and joy-filled memories of my dad blew away the lingering resentment I had towards God in that moment. At the same time, Brad started humming the tune. I stopped laughing mid snort and just stared at him.

"What?" he asked.

"What song was that you were humming?"

"Huh? Oh, I didn't realize I'd hummed it out loud actually. It's... wait a sec. I know it, I just can't think of the name right now but it just popped into my head... what's wrong?"

My face must have shown my shock. "Nothing's wrong, it's just I literally had that song pop in my head too. And you know I mean literally when I say it."

Brad raised a brow. "Well, with everything going on between us I'm not all that surprised. I've actually been thinking that you must be developing a sort of enhanced consciousness from all this that perhaps I'm picking up on. What's that song called anyway? I know I know it."

"'Heaven is a Place on Earth.' Belinda Carlie, from way back in the pre-millenium 80s."

Brad smiled. "How appropriate. Plus, music is a creative gift of this world that God definitely speaks through as well."

That gave me pause—God speaking through our creative gifts? I snatched up my phone and started typing notes. "Well, there's a story there for sure."

"There's a story behind every door,." Brad mused.

I finished my flurry of notes, elated from the download of creativity. "Thank you for that." I gave him a kiss and gestured to the digital brain, eager for more creation. "So, back to our not-so-regularly scheduled programming. What does this circuit do?"

"Now you're reading my mind," Brad grinned as he picked up a tablet and typed a few lines of code like an expert pianist. The monitors before us switched over to show the digital brain displayed as binary data mixed with what looked like mystical mist. It looked like a digital spiral galaxy.

"René Descartes described the pineal gland as the "principal seat of the soul," and believed it to be the place where all our thoughts are formed. As I mentioned, what you described from your 5D experience with Charles re-sparked something I actually already knew, but got buried along the way: What if the pineal gland is the portal to consciousness—but not just any consciousness; the collective consciousness?"

He paused, letting that hit me. The meaning of his words resonated through me. "I feel the truth of that...but there's more to it."

"There's always more." He nodded to the digital spiral galaxy in front of me. "Take a closer look."

I stepped up to the monitor and gasped. What my eyes had interpreted as mystical mist from afar was in fact made up of micro-sized spirals of code, too.

I took another step forward, entranced. Suddenly, the image on the screen glitched and the digital spiral expanded, glowing brighter than before.

I glanced at Brad who was smiling like he just won the lottery.

"I knew it. Your energy amplifies it in the same way our pineal glands are affected by light and darkness. The circuit responds to your 5D-enhanced energy."

"What does that mean for Uncle Jeff?"

"If my calculations are right, which, let's face it, they usually are, if I upload Jefferson's digital essence using the Pineal Portal Transfer Algorithm I came up with, Jefferson Sands could be back online with us as of tomorrow. The energetic essence we captured through the digital mapping of his connectome is already infused with your 5D-enhanced energy. It should work, Samantha. I'd stake my life on it, actually." He side-eyed me as I marveled at it all. "I wouldn't stake yours on this one though, if that factors into your decision at all."

I looked at my feet.

Did that mean he loves me?

Ugh, so not the time for the romantic melodrama, Sam! Focus. You're literally in the middle of the evolutionary breakthrough you ignited!

"My decision, eh?" I replied, pushing away the noise.

"Of course. Anything we do with Jefferson's essence and this bot are entirely up to you. But imagine where all this could take us— and take the world to? This opens doorways of potential not just into digital resurrection but digitally enhanced healing, anti-aging, knowledge expansion..."

He was pacing the room now in that way he did when he was so into an idea he couldn't not move with it. I loved what he was saying. Loved his energy. Loved...him?

The latter I couldn't put a period to just yet.

Wouldn't I be able to if this were meant to be?

I remained silent in my ruminations as Brad kept going on and on. His obsessive pacing around me and the table we'd fucked on so many times now seemed to imprison me. His relentless enthusiasm for all the potentialities of the Lazarus Online project was closing in around me—

Did he want me only because of what I brought to this project?

I caught my downward spiral and forced myself to break free. I squared my shoulders, took a deep breath, and strode over to Uncle Jeff's digital box.

I laid my hand on it, feeling the vibration of his essence through the steel machinery. The data on the monitors the box was hooked up to streamed faster as I did so. I smiled.

I really had something extraordinary, didn't I?

I walked the line to face AI-bot-Jeff. I took its hands, feeling how real they felt. The spiral galaxy displayed behind it spun faster, the digital mist sparkling around and through it. I placed my hands where the bot's heart would be, feeling nothing. An inner hollowness made itself known to me then; a hollowness I needed to fill lest I die in the suffocation of it.

The digital spiral galaxy slowed and dulled in brightness then. I quirked a brow at that but had too much in my mind to hold anything else at that moment.

I turned to Brad. "Let's bring Uncle Jeff back online."

CHAPTER 22

The rest of the day and into the evening was filled with more of Brad's elation and exuberance for the infinite possibilities the Lazarus Online Project would bring to the world. And as much as I tried to keep up, since my initial elation in my decision to move forward with bringing Uncle Jeff back, something inside me was so not on board. I went along with Brad's excitement though, hoping that would authentically respark it in me. I felt it myself, but in a strange, muted way, as if I was dissociated from joy.

What the fuck is wrong with me?

That incessant thought pounded upon my brain in a reverberating rhythm from many a nightmare.

By the time evening rolled around I really wanted a drink—I'd even go so far as to say I needed one.

Brad and I were at my place, a rare occurrence in the four months we'd been secretly seeing each other. I knew this came from my ingrained patterns of choosing to stop by, chill, hang—whatever we ended up euphemizing it as—at whatever guy's place I was seeing so I had the freedom to escape when I wanted.

In my heart of hearts, I didn't want to escape from Brad. I wanted to enjoy this while it lasted like basking in sunlight on my personal, private beach. But the dichotomous dance was alive and well within me, swinging in my darkness to shadow my light time and time again.

I couldn't follow Brad's train of thought as he went on and on about Lazarus Online. I used my breath to help me break free of this overwhelming funk yet again. My Merkaba spun back into alignment as best it could, but it felt as heavy and burdened as I did.

At least now I could hear and somewhat follow Brad's excitement.

"... what if we even get to the stage where we can extend one's consciousness into something like, let's say..." He looked around as if about to pluck the answer from thin air.

Maybe that's how it was for him? The creative downloads I got usually shook through me like lightning strikes. But for the laid-back types like him, it made sense that they would flow in like a breeze.

Damn him, he made ease look so easy.

"Oh wow—I just had a flash of something here... what if the Digital Pineal Gland and the Pineal Portal Transfer Algorithm combined could be a way for humanity to literally explore the heavenly realms of the fifth dimension? Similar to how you do but connecting consciously to... Oh, wow, what about this: Connecting consciously to Christ on a level we've never seen before." He paused for a moment, smiling at something I could tell was a bygone memory. Then he closed his eyes and recited:

"Very truly I tell you, whoever believes in me will do the works I have been doing, and they will do even greater things than these, because I am going to the Father."

He opened his eyes, alight with something brighter than I'd ever seen before. "That was my dad's favorite scripture and one he left on my heart. Samantha, Jesus tells us here that if we believe in him we can not only do what he did but literally do greater! What if Lazarus Online is our greater work that begins our build of God's Kingdom of Heaven on Earth?"

As he had been speaking, that darkness that had swung in before had been lifting, softening me. I felt lighter, freer, capable of receiving what Brad was saying. There was a presence—even more than that really—there was an atmosphere around him that I couldn't deny. "Jesus, eh?" I mused. "Honestly Brad... I'd never understood it like that before—"

I broke off, seeing my phone light up, no sound due to the silent mode Cindy suggested for after work hours. OK, no, she'd actually snatched my phone away and set it up herself. I thanked her silently as I forced myself to ignore it, but that familiar dark burdensomeness started to pull me into a spiral of wondering who I was ignoring and I suddenly felt overwhelmed to an intense degree by everything I'd just heard from Brad.

Which is of course when he started up again. "I'm so glad to hear you say that. It truly is in how you receive the word and who you receive it from that counts. I can show you so many scriptures that I know would light up that beautiful mind of yours. They're honestly more like codes than anything. Hence why coding has come so naturally to me." He whipped out his phone. "Let me find a couple I just know you're going to love."

Ugh no, shut the fuck up already! Stop assuming you know what I love, you railroading fuckface!

The darkness had latched onto my overwhelm and my inner angst was spiraling. I'd fallen out of whatever understanding had just shot through me, leaving me feeling bereft and alone in a sea where Brad was speaking some other language of light and joy at me I couldn't comprehend.

Aloud, I tried to redirect the flow of the evening as I picked up the remote to search for something to relax and watch. "There's actually an old series about Jesus I'd heard was great—"

In the corner of my eye, I saw my phone FLASH—FLASH—FLASH—

Fuck off, whoever you are!

But that was three flashes.

Shit, bad things come in threes.

I abandoned the remote for my phone, still determined to have my chill evening. Distracted, I tacked on, "—want to watch it?"

My ask came out as an inaudible mumble and Brad kept talking. "Yes! The Chosen! Let's see if we can find it. You're going to love it! You know, I actually believe you've been chosen to wield this gift of yours..."

I was barely listening as I read the novel of texts that had built up from Chairman Ross. They started small and grew with each message. It was like I was climbing downstairs into the basement of his breakdown.

A breakdown I had ignited.

It was the one that said, "What have you done to me, you bitch of a headcase?" that evoked The Scream.

The Scream that tore through my throat from the bowels of me— an ear piercer that wrenched my soul and ricocheted within me like a caged animal from my haunted past.

At least it got Brad to shut the fuck up—

—Only to rush right over to me. "What happened? Are you OK? Do we need to go to the lab? The hospital? I can call my doctor."

His touch felt like a cloying cage. I couldn't take it and pushed him away.

"I'm fine! Just leave me alone!" That was actually the last thing I wanted, but that Suspicious Sister inside me definitely did. Or at least, she thought she did.

I felt a fissure form in my heart as she rose up within me. A fissure that became a fortress.

I stormed past him to barge out the door when—

"Fuck me, this is my place!" I froze and slowly turned toward him. "Now this is something I've always wanted to say—get out!"

It's almost like he was ready for that. "I trust that's not the real you talking, and I'm here for whatever parts of you need to process right now." His sympathetic eyes nearly lulled me into that calm flow of his I had envied just a moment ago.

But this Sister was having none of it. I tried to take the wheel back, but she held on like a vice with a vicious streak, careening us into a head-on collision with my heart.

"Oh my fucking God, don't use your psychology Ph.D. bullshit on me. Get the fuck out! This was a mistake, it's been a mistake all along, why the fuck did I even think this was—"

Suddenly, he was in front of me, like a pillar of sunlight against my despairing darkness. I had a surreal moment of déjà vu then, echoing so many other times I'd had an episode like this—with past lovers who'd gotten too close, Uncle Jeff, even...

Fuck me, this had even happened with my dad. Something in my soul would absolutely not let any man, let alone Brad, get closer. She'd been triggered out from the shadowed depths of me, ready to take down anyone and everyone.

Brad was her bullseye.

And he took it like a champ.

The words had hurt him, I could tell, but it was like he was seeing through them once again... to the subtextual seeking of help at the heart of this self's terror.

My Suspicious Sister kept slinging the hate as he enveloped me in his arms. I felt my body thrash and writhe to escape what I was coming to realize was my mind and body's misinterpretation of comfort as confinement.

Something had happened to this self... something I couldn't see...

She wouldn't let me.

But she also wouldn't let anyone help me.

NO ONE CAN SEE NO ONE CAN SEE!

Her all-pervading mantra pounded in my mind as she fought like a mad woman to push Brad away.

It felt like I was going to rip in two as our dichotomous dance turned into a tug of war, the winner of which kept her love.

I saw her then, in my mind's eye. It was like time stood still within me as we glared at each other along the line of the rope, like enemy pirates on opposing missions.

Suddenly, her form glitched and shifted—allowing me to see beneath the glare...

To a fractured child...

A bygone me.

It was her energy that had kept me on edge all these years, but she'd also steered me away from plenty of clear-and-present-danger-type scenarios.

A foe who was also a friend.

An enemy who was also a sister.

A shadow I could invite into the light.

My light...

Our light?

I surrendered to it.

She did not.

A fragmented transcendence enveloped us—

And I let it take me.

CHAPTER 23

I woke up with a start several times in bed that night, Brad at my side to soothe me back to sleep. Around the fifth time I stayed under, falling into a dream I'd known was coming all along...

I was running through the window-lined hallway of my childhood home once again. The space had a muted sepia tone as if aged with time.

I was in the body of my child self, tiny hands reaching out to swat away the billowing curtains, blocking my view of the hall ahead.

I took a backseat in the corner of her mind and let the scene play out.

"Ready or not, here I come!" I felt her speak the words, but this time they were hollow, echoing a tone of hopelessness all around us.

"Ready or not..."   "Ready or not..."   "here I come..."

"Ready or not..."   "Ready or not..."

"Here I come..."

She froze, clamping her hands around her ears, trying so hard to shield herself from her own suffering.

"I give up, Uncle Jeff, okay? Come out, come out wherever you are!" Her poor voice broke on the last word as she crumbled to the floor in a huddled heap like a tiny turtle. Her small form shuddered, then she dared to peek out and looked around, removing her hands enough to hear—

"He's never coming back for you—"

"He left you—"

"Rejected you—"

"Abandoned you—"

The cacophony of her own worst fears echoed harshly all around her. It was like being inside a clanging bell of discordant noise.

"NO! Come out, come out, Uncle Jeff—PLEASE!"

It was the "please" that did it.

The word echoed from within and resounded through us.

With my adult awareness, I helped my child self calm and rise from the floor as the noise died down and Uncle Jeff strode through the curtains that fluttered around him to stillness like stalling sails.

My child self and I saw him as one, through a clouded lens of tears that streamed down our face as we ran to him.

I felt us growing and transforming into my adult self as we ran, that child me healing and integrating as I engulfed Uncle Jeff in a hug, imbued with all the love I had for him.

"I knew I'd find you. I trusted and I knew! This is a sign you're ready to come back, isn't it? Oh my God, Uncle Jeff, just wait until you see—"

He broke away from our embrace in the gentlest of ways and stood back to really take me in.

"Oh Samsey, your story-wired mind is a wonder to behold. Now that it's paired with your expanding energetic consciousness, even a spirit such as I can't rightly see where it'll take you. But it's up for sure."

I sagged in relief at his words. "It already has in so many ways, but there's still so much weighing me down."

Uncle Jeff chuckled. "Of course there is. The Other, the Shadow, the Darkness, the Ego. All these names we've come to call it only further separates us, when unity is what we really need."

I thought back to my tug-of-war with my Suspicious Sister and shuddered, feeling how many more there were to go. But that was for future me, now I had my Uncle Jeff before me and I wasn't going to waste a second of it. "Unity is the next step in your digital resurrection! I can't wait to see you in—"

"That's not what I want, Samantha."

I froze, elation obliterated.

The only time he'd ever called me Samantha was when I refused to get out of bed for a month after Dad died.

"But... I thought... I mean—isn't that what these dreams have been about? Isn't that why you're here?"

He softened, brushing stray hairs out of my face. "Your ability to find meaning in so much is one of your greatest strengths, but in this case, no. I'm here because I can't leave."

He locked eyes with me then, saying more in those vibrant green orbs than words ever could.

He's lost in a limbo you trapped him in by trying to keep him. You weak, selfish bitch.

I gasped as that realization hit me like an icy fist, punching me in the gut.

"None of us can go against the Divine Plan, Samsey. And even if we can, it doesn't mean we should."

I crumbled into his arms, sensing the truth he'd spoken as a lamenting loss echoed in the hollowness of me.

The dream whirled around us. I instinctively gripped onto him, holding onto these last moments.

"I'll be okay, Samsey, and so will you." His words whispered through me as I released him and the whole world fell away as I let the darkness consume me.

***

I woke up on a pillow sodden with tears, Brad asleep in my bed beside me. I brushed a stray hair from his face and in that instant realized how much love I had for him.

Whether I was in love or not, I still couldn't tell.

"I'm sorry," I whispered. I went into stealth mode and tip-toed out the door, grabbing a steel baseball bat that I kept by the door as silently as possible as I did so.

CHAPTER 24

The Digital Dungeon doors opened with a swish as I slipped in, feeling like some kind of cat burglar. The data streaming on the screens illuminated the way just enough for me to see AI-bot-Uncle-Jeff looming before the table where my shiny metallic target awaited.

I looked the bot in its gray-green, marble-like eyes, repulsed at how fake they looked compared to Uncle Jeff's vibrant green eyes I'd just seen in dreamland. Those eyes filled with so much pain, the pain of being trapped. I stared into those dead eyes as—

SMASH!

Steel met steel as I brought my baseball bat down on his digital reliquary.

Something dark and feral overtook me as I kept going.

SMASH SMASH SMASH—the Hulk had nothing on me.

Suddenly, I felt a cloud of trapped energy burst forth from the box. I saw it as a silver-blue cloud, not just in my mind's eye anymore, but right in front of me.

I knew it was Uncle Jeff's spirit, and felt the loss of him as the cloud beamed up and dissipated from sight and sense.

Stupid, stupid, STUPID!

What the fuck did I think I was doing messing with the Divine Plan?

Who was I to think I could do anything of consequence...

Who was I...?

A ringing resounded within me, an alarm of anguish that came out as The Scream 2.0. I took it out on the only thing I could at that moment...

AI-bot-Uncle-Jeff.

The ringing intensified as I smashed to pieces the manifestation of my clinging need for my Uncle, which echoed back to clinging to my dad's memory, clinging to the thought of what could be or could have been with every lover yet running when it felt like "too much."

What the fuck is wrong with me? What the fuck is wrong with me? What the fuck is wrong with me? What the fuck is wrong with me?

The compounding cacophony of the thought that had always haunted me rose up like a wall against what was truly under it...

You're just not good enough.

There it was in black and white—literally, that's how I saw it in my mind.

Black bold words on a stark blank page.

I wobbled, weak and woozy, steadying myself on the bat like it was a cane. I giggled at the feeling of it all—like playing that stupid drinking game, dizzy bat, but sober and no fun because of it.

That's when Brad walked in.

I must have looked like a crazed kid in a candy store with a vendetta against sugar.

I was woozy as fuck—akin to the drunkness that I thought would alleviate pain, but instead brought it way too sharply into focus.

"He's free," was all I could say as I fainted upon the wreckage that had caused Uncle Jeff's purgatory.

I'd forgotten that it was also a marvel Brad and I had created... together.

All of a sudden, he was right there, catching my fainting form.

I felt the full weight of my regret and grief as his eyes bore into mine and the world went dark.

CHAPTER 25

I floated through dark ether in the recesses of my mind.

It was nice here. I didn't want to leave. I wanted to stay hidden here for all eternity. I snuggled into the darkness as if it were a warm blanket I could hide under like a child who didn't want to go to school. It had a semblance of bliss to it, and the semblance was all I needed—

"Who are you?"

"Ugh, leave me alone."

"Who are you?"

"Fuck off!" I snuggled in harder, clenching my eyes closed. My body went rigid.

"Who are you? Who are you? Who are you? Who are you?"

"I DON'T FUCKING KNOW!!!"

And yet, as soon as I yelled that—I did know.

The interrogating voice ceased and, from the silence, I felt more than heard the gentle laughter of a woman. Somehow, I already knew who she was.

"Not knowing is the first lesson, Samantha Sands. Congratulations."

Suddenly, the darkness dropped, and I was engulfed by a light so pure I felt scalded by it.

The light expanded and morphed, becoming a giant torrent I was moving through faster than anything I'd ever experienced.

I felt myself ascend...

Bend...

Transcend...

***

When I finally came back to myself, I was able to see with new eyes that looked upon the world with a clear sense of knowing—while knowing nothing at all.

My newly embodied bliss was thanks to the woman seated before me—Shiba. The one I'd abjectly refused to see was, of course, the one I ultimately needed.

She was an ageless, wise wonder. Long dark hair streaked with finely aged silver. Her eyes like windows into worlds beyond understanding. Her presence like a nurturing mother and a strict teacher all in one.

"You have learned much in our time together, Samantha Sands."

We sat together in a vast green garden, the birds in the many trees surrounding Shiba's state-of-the-art ashram singing as she spoke. Their peaceful tones resonated with each other, creating a new note that struck my soul with its magnified power.

"Now it is time for your reflection ritual," she continued, "a sacred process to integrate and prepare for the journey back to your life."

I felt a far-away part of myself tense up at the thought of returning to the mess I'd left behind, but all fear was met instantly by a poignant peace.

"Thank you, Shiba." I clasped my hands and bowed, hearing the birds now echoing my tone.

CHAPTER 26

S hiba led me through the labyrinthine halls of her ashram, and when I said this was state-of-the-art, I meant it. The place was more like a luxurious compound. I'd been told it was built from pieces of the holiest temples and shrines around the world that Shiba had taught at. Carved marble from Greece met with warm wood from India, woven with bamboo grain from Japan, and aged stone from Egypt.

I felt the harmony of all the pieces synchronizing together. The vibration I had ascended to had given me the ability to perceive the energetic interweavings of magic, memories, and healing that had soaked into it all over the centuries.

We came to a door then, carved from aromatic sandalwood and inlaid with the intricate pattern of the flower of life. The delicate design seemed to pulse with a quiet wisdom.

Despite the serenity Shiba's Sanctuary (as I'd dubbed it) evoked, I felt a twinge of fear at the energy emanating from the door. I knew I'd have to feel back into who I used to be and realize things I didn't want to.

Would I be able to come back to this ascendant level?

Did I deserve to?

Shiba's eyes flicked to mine, perceiving that lower-level thought.

"You are ready, Samantha. Do not fear the doubts; let them become keys to all the doors you were once afraid to walk through."

As I held her gaze and her words flowed through me, the resistance that had risen cracked open like windows within me. I nodded and bowed, opening the door.

Inside was a floatation pod made of shimmering metal, inlaid with tiles carved with spiritual symbols reflected in the still water.

I felt like I was stepping into a coffin as I disrobed and lay in the warm, shallow water.

"When will you come back for me?" I asked.

"When you know you're worthy." She replied simply.

I caught her signature smile she had for me. She had one for each of her students. In her gaze I saw the highest self I could be, even beyond what I'd already achieved.

Why could I STILL not fully see that for myself?

That was my last thought as Shiba closed the reflection pod and I felt the pull of introspection like a drain within me, swirling me down, down, down, and then—

IN!

Into a memory, as if dropping in on a stage scene like a Cirque du Soleil performer on wires—without the wires. It was similar to when I was force-ghosting in my memories with my younger selves, but here I was above it all. The watcher of my woes.

I saw the scene below me play out...

It was two months ago, the moment I'd arrived at Shiba's and, spoiler alert, it was anything but graceful.

In the memory, Cindy's Range Rover pulled up to Shiba's palatial ashram. As I watched, I recalled we'd had a heated conversation—

The instant I recalled it, I found myself in the vehicle.

Cindy and Arriving Me in the front seats mid-heated convo:

"...and make sure to check in bi-weekly with Chairman Ross, but let him be as much as you can aside from that until I get back. IRIS has everything loaded up and scheduled with the marketing materials and budget reports. Brad—uh–um...the Falls Tech team is underway with the IRIS Creative Support Fleet. That should all be ready for me to go over when I return. Then we can get back to yet another new normal."

I watched Arriving Me, seeing through her carefully laid plans and scheduled-out agenda to the fear that was really underneath it.

"Sam..." Cindy hedged, carefully choosing her next words. "Remember, it's just a company. Albeit the best I've ever worked for, and soon-to-be the most evolved in the Creative AI Blend Space, but it's not worth you losing it over."

"I haven't lost it, I'm..." she paused, looking out the window at the ashram. Then she scoffed. "I'm wandering."

Cindy smiled. "Let yourself wander, Sam. We'll be ready for your 'Return of the Queen' moment."

"Thanks, Cindy." Arriving Me was about to get out when—

"Wait, Sam," She held out a letter. "Brad, umm... he gave me this to give you. He really wanted to drive you. He—"

"I know he did, OK!?"

Woah—this I didn't remember, but it came back to me in full force as I saw the vicious about-face Arriving Me had just performed. She snatched the letter from Cindy's timid fingers, wrenched open the door, and slammed it shut in her face.

I felt the pull to Arriving Me as the memory began to dissolve, but not before I saw the hurt in Cindy's eyes. The hurt I'd caused. The memory started swirling me away...

NO! WAIT—

And suddenly, everything did wait...

The dissolving haze hung in the air.

Cindy was frozen in time.

Arriving Me was also frozen, mid-stomp, on her way to the ashram.

I was free to move within it, though.

Suspended in this strange time-space, I felt the hurt I'd caused Cindy. I didn't want to, but I allowed myself to feel it anyway. As I did I visualized a double-pan balance scale—the pain in one pan, acceptance in the other. At first the pain, looking like a cluster of dark malignancy, weighed the shining apparatus to its side, but as I accepted what I'd caused, white light accumulated in the opposite pan, and the scale tipped, rebalancing.

Cindy and I let out simultaneous sighs of relief, sighs I saw as stagnant shadowy energy leaving us both. I realized then that I'd healed us from what Arriving Me had caused. I didn't know how exactly, surely Brad would have some quantum philosophical theory about it, but I didn't need to know the how. I just knew that it was done.

I let the memory swirl me away, Cindy's smile at Arriving Me's angry back ushered me forward, into another memory...

Of my first meeting with Shiba.

Arriving Me sat in the garden, scowling around at everything. This one I remembered, and knew she'd shoved Brad's letter down her bra—that and her underwear being the only things she was permitted to keep on of her own underneath the gray garb Shiba's attendants had given her.

They'd called it a kasaya. I'd called it hideous.

I giggled, remembering my recalcitrance as I floated above the scene.

Shiba entered, wearing her own white kasaya. I saw Arriving Me narrow her eyes at that, comparing and feeling inferior.

Wow, how much energy I used to waste on that toxic seesaw!

As I thought that, accepting the way I used to be, I felt a great weight leave me.

In the memory below, Shiba glanced up with her signature smile for me. She saw me. I knew beyond the shadow of any doubt that would dare to try me on it. I smiled back with a bow, knowing I'd integrated the meaning of this moment, and with that the scene dissolved into...

A meditation room made of wood that gleamed like gold. I could smell the earthy cedar of it as I watched the Me who was then a novice sit cross-legged, facing Shiba. I vaguely remembered this one. She'd facilitated so many regressions that it all felt like one big never-ending story.

We both had our eyes closed. Shiba looked like she'd attained Nirvana as usual, whereas Novice Me had her eyes squeezed shut, tenser than a sinner on Sunday. I could feel waves of angst emanating from her like self-inflicted poison.

Yeesh, was that how I used to present myself to the world?

I caught myself in my own self-judgment and let that come up on my inner scale to be released and rebalanced.

But Novice Me below was still so full of it. So caught up in thinking that she should know better by now, calculating and comparing herself into ruin.

She let out a frustrated sigh, clenched her fists, and scrunched her eyes shut even more, trying so hard to get whatever it was Shiba was teaching.

"Samantha," Shiba said gently. "Let us pause for—"

"No! Wait a sec, I've almost got it."

"You will get it, there is no—"

"Yes, there is a rush! I have to get back to my life!" Novice Me shouted, eyes snapping open in fury. "Maybe you have the time to lounge about in your palace, but I don't! I need to get back to work!" She'd escalated to yelling while Shiba remained calm; fluid like water, still as a statue.

Novice Me noticed the contrast and finally slowed down, took a deep breath, and let herself fall back onto the floor.

"I'm sorry, Shiba. My Higher Self is probably rolling her eyes on her pedestal up there, eh?"

I laughed at that, leaning in. This was like watching a movie I'd already seen that was unlocking truths and realizations in me from this higher vantage point.

Shiba smirked as her eyes flicked right to me, floating above them. Just a quick acknowledgment before she laid down beside Novice Me.

"There's a reason for this, Samantha. There always is. Something in yourself is blocking you from getting to the memory you need to feel and heal. Where do you feel the resistance most?"

"I don't know, but it feels like I've got the worst heartburn along with the worst hangover ever."

"Your mind and heart are still out of alignment. The key to your quantum abilities is the coordination between them."

"I know that, but—"

"Do you know? Do you truly?"

Novice Me scowled. I could feel how much she wanted to rebut with a sarcastic quip, but instead, she breathed it out.

"No...if I did, there'd be no resistance, would there?"

Shiba smiled and took Novice Me's hand. "You are ready, Samantha."

Novice Me said nothing as she gripped Shiba's hand in gratitude. A single tear fell down her cheek as she closed her eyes and let the memory that needed healing pull her to it—

—I felt it, too, and I was pulled back with her, but in a detached way as if the two of us were connected with a flexible, ethereal umbilical cord. I went with it, along for the ride as we were both pulled back through a swirling abyss and into...

Another memory. A memory within a memory—Chris Nolan's got nothing on this.

Novice Me floated near the Me in the memory, force-ghosting away. I floated above it all, watching the scene like an all-seeing audience.

It was a memory of myself and my head writer, Michael Strassner. We were in the writers' room at Sands Studios, just the two of us. Memory Me hemmed and hawed before a huge whiteboard covered with a rainbow of index cards, outlining the show we were discussing.

I was strangely drawn to Michael's calm emotional tone as he analyzed the board, as opposed to Memory Me, who was definitely overanalyzing. I could feel the difference between her and Michael's energy. Michael's created a space for me to be part of his process; Memory Me's felt like a closed door to collaboration. That realization struck me—a harsh hit to my heart.

I hadn't intended for that to be my vibe, I just wanted...

Oh wow—the word caught like a lump in my throat—

I just wanted it to be perfect.

Suddenly, Michael's eyes lit up, as did his energy. I felt it like a spark of divine inspiration in his mind. The power of his imagination was something I'd never fully grasped or appreciated until now.

"Oh my good God and all that is holy!" He pointed to a card. "Sam! What if we move that to the mid-season plot twist and combine it with Baratheon's fight with Chief Santor? That'll create a double whammy mid-seasoner that can set up a doubleheader climax when Syeida and Baratheon join forces with the power of the bloodstone backing them!"

Memory Me's eyebrows furrowed, her energy pulling me back to float above her and I saw Novice Me force-ghosting beside her. I remembered this now from Novice Me's vantage point, knowing she was feeling and accepting Memory Me's does-not-compute distraughtness at the thought of her carefully laid-out story crumbling under the adjustments Michael was proposing.

What they both were missing was Michael's heartfelt ingeniousness that was pouring out of him.

"See, if we do that, we can revise the rest of the season so both

Syeida and Baratheon have more equal show time per episode till the season finale." He was up and at the board now, drawing a tiny sketch of his idea, very obviously showing that he wasn't intruding on Memory Me's ideas.

Memory Me, however, didn't see it that way. She managed a smile through the cringe that resulted from her overthinking the adjustments needed. How one plot point change rippled out like a boisterous butterfly effect on the whole. The story she'd worked so hard on, her baby, who was "perfect" just the way it was.

"That's... ummm... that's a great idea, Michael. I love how you're thinking! Thank you for so thoroughly considering their characterization."

She paused, finger-tapping her lip as if she was really considering it, but I knew better. Not only had I been her, I'd been force-ghosting beside her in this memory as Novice Me, feeling her dismissing Michael underneath her pretense. I'd forgiven myself in that regression but hadn't sensed into how my actions had affected Michael.

I felt that loud and clear now as Memory Me wiped off his squished, neglected sketch. "But this is Syeida's story, not Baratheon's. He'll get more showtime next season. Don't worry, I've got that all planned out, too."

Michael smiled through his disappointment.

Ooof!

I felt that like a stone weighing down my heart.

"Of course, Sam. It was just a silly, spur-of-the-moment idea. Couldn't let it get away without seeing if it would fit."

He picked up his papers and bag. "Seems like you've got everything worked out. Anything you need me to stick around for?"

Memory Me had already gone back to overanalyzing the board. "Huh? Oh, no, I'm all good. You head home, I'm sure Alan misses you. Thanks again for all your help this week."

I saw the moment where he wanted to say more but held back. "Anytime, Sam."

He shuffled out. I saw Novice Me dissolve away... I began to dissolve too, but something in me wanted to stay and follow Michael.

"Wait," I commanded calmly, remembering how I'd done this in the memory with Cindy. The memory didn't freeze like it did last time, it kept going, but this time I followed Michael out...

Out of my memory of this moment... and into his.

I suddenly realized that's what I'd been experiencing this whole time—his experience of me.

He walked down the "writers' room walk," as we'd dubbed the extra-long corridor that connected all the writer's individual offices and communal "breaking story room." He slipped into his office as his phone buzzed.

"Hi Alan... No... I mean yes, I'm all done and heading home... I did. But...um yeah, it's better as it is. Sam's structure is always on point. I shouldn't have wasted my time with that idea or yours by practicing pitching it. Sorry about that, honey..."

There was a long pause. I could only hear a muffled man's voice—but he did not sound happy. "Aww hon, it's not like that. Sam's a genius..."

Whatever Alan said on the other end brought Michael to tears, most of which he swallowed. "I love you. See you soon."

He hung up with a sigh as he leaned back against his closed door for a few breaths. I felt him pushing down so much repressed disappointment—in himself, his ideas, his own imagination.

A tear rolled down my cheek as he walled it all away and shuffled out. Business as usual.

I tried to hold the memory and heal what I'd caused in him, but it dissolved away, leaving me bereft in a sea of regret.

CHAPTER 27

I let the reflection ritual process take me, and found myself floating through a gray somber haze that pushed me into a driven reverie.

Was that how I'd come across to Michael for all these years? To the whole room?

Fuck me, I'm an asshole.

I couldn't help the thought.

I felt it rattle around in my mind, and gasped at the perplexing pleasure that arose from my abhorrent discomfort at realizing that I can, in fact, be an asshole...

But that doesn't make me one. Not fully anyway.

That thought made me laugh out loud, considering all the percentiles of asshole one could be.

The soapbox speech I'd declared to the board so long ago streamed through my mind like a harpy's cry:

"Why are we watching shows if not to connect with the human spirit who created them?"

I felt the hypocrisy from the version of myself who'd spewed that float through me. I saw it leave my body as a shadowy murk, becoming one with the gray haze that now suddenly became dense. An ever-pressing weight I could not move through.

I'd learned a few tricks during my sojourn with Shiba, though, so I knew, like a precise slice of knowledge cut just for me, that if I surrendered all would be well. It took me another minute, but when I did the weight shifted, and the haze began to dissolve...

This dissolving was different from the others. It swirled with the shadowy murk that hadn't cleared, and I traveled through it with a jerking motion, not the smooth sailing that had come before.

Suddenly, the jerking became rhythmic driving thrusts of pleasure and I exploded into a memory of me and Brad—

He was fucking the Me in the memory. Hard, fast, rough. All the ways I'd said I liked it. Our sexual connection was so strong I felt the thrusts and anticipatory orgasms within Memory Me from my vantage point.

Novice Me was there too. Force-ghosting next to Memory Me and obviously way too ingrained in it all to have more perspective.

"Fuck me, Professor. Fuck me harder!" Memory Me's plaintive sexual cries resounded in the space. Thank God Brad's brownstone was on the end of the row. Any wall-sharing neighbors would definitely have had some noise complaints.

Brad did what Memory Me had asked and then some. I could tell they were both close to climaxing as he held off in that impressive way he was somehow able to while Memory Me came, the convulsions deep inside her setting off Brad's orgasm as well. She was splayed on his desk, eyes clamped shut in bliss, edged with pleasurable pain.

"Look at me," he commanded. She did but kept looking away.

"Look at me," he repeated. "Don't take your eyes off me."

I watched Memory Me as she did. I'd realized in the first regression of this memory how much it had taken to push through the resistance that wanted her to stay hidden from this man who I could now sense was doing all he could to be with her—with me...

Their eyes locked in a moment ripe with tantric heat...

That exploded on a peak of pleasure as they both cried out and sagged into the weight of each other, limp in the languor of lovemaking.

I kinda thought that might be it and was preparing for the memory to dissolve me away...

But it didn't.

Brad gave Memory Me a kiss as he got up and disposed of the condom. He handed her a blanket from his office couch and got dressed. She took it gratefully with a smile—but it was a smile cracking with way too much overthinking going on behind her eyes.

"Write me well if you publish a book on this. Ah, I've got the title: *Sexual Kink 101*. Ha! Bet you'd get a lineup of students wanting just what you gave me. Well played, Professor Falls. Well played. I'll help you write the perfect pseudonym for me."

She got up and put on her clothes as Brad watched her. The energy of the memory shifted over to him fully, as it had with Michael, and I felt him *see* into Memory Me, in his own energetic way, beneath what she was saying—

To the insecurities her sarcasm was trying to hide.

"I bet we could even make sex bots eventually, imbued with all of this. Sexual inspiration knows no bounds, eh?" she remarked, her flippancy a flimsy foil.

"Do you really think that's all this is to me?"

Memory Me slowed down at that, but quite literally couldn't stop moving around the room; fidgeting, fussing with the furniture, and making like she was doing her hair in a mirror.

"What? Of course not. I'm just kidding, silly."

"You wear the mask of sarcasm well."

"Hey! I'm not—wait, that was good." She whipped out her phone and typed it into her notes.

Wow. Watching this play out from my watcher vantage point I saw exactly what Brad was trying to point out to her. As he watched Memory Me, I could sense how much he loved her... loved me...

I felt beneath his exasperation with my avoidance to his admiration of the genuine artist within.

Memory Me caught Brad's eye as she finished typing. "What?"

Brad just smiled. "I love you."

It wasn't the first time he'd said it, but those words caused Memory Me to freeze up under his gaze. Brad felt it too and took hold of her hand, using the other to guide her eyes to look at him.

She did. I remembered this eye lock. She was lost in his eyes, as if suspended on a precipice of a before and after that began with those three words.

"Oh, Brad. I love so much about you. I love our time together. I love what we create. I love... so much..."

"But?"

"But I don't know if love..." she gestured between them. "This kind of love is possible for me."

There, she'd said it. Admitted a fear she'd long since carried. I could feel her waiting for him to leave. Our inner saboteur wanting him too.

Instead, he kissed her forehead, breathing in her scent.

"Anything's possible, Samantha. Just depends on what you're willing to fight for."

They locked eyes again. I remembered this moment as I had experienced it while watching it like a reverberating déjà vu, all the suspicions and calculations she was doing in that beautiful mind that was both our greatest gift and crippling curse.

A mind that was so hell-bent on emotional survival it walled out the very thing that would heal it.

Brad knew it too, I could feel his energy wanting to help me out of this resistance without being too pushy. He believed in us that much—

BUZZ—BUZZ—BEEP—BEEP

"Oh, for fuck's sake," Memory Me cursed while retrieving her phone and pushing down everything she was just feeling.

Brad smirked, knowing she was about to say...

"Sorry, I've gotta—"

"I know you do." He squeezed her hand as she pulled away.

As the watcher I saw the pain in his eyes, felt the longing in his heart as his fingers lingered on hers...

Then he turned away and left the room.

A sudden urge came over me, one I'd never experienced. One I'd never wanted to.

I longed to go with him.

It was as simple as that.

IRIS, Lazarus Online, Sands Studios—none of it mattered in that moment when all I wanted was to be with him once more...

I ran to the door, but I could not pass. I rattled the knob—pounded on it. All to no avail. I turned to Memory Me taking the call.

"You fucking idiot! Do you not see what you're doing to him? Do you not see what you're letting... walk away?"

I floated down from my raging high horse, realizing who I was yelling at. Then the realizations of all the pain I'd been the cause of in these regressions reverberated through me.

It felt like an echoing scream of torment that would last forever.

The memory jerked and swirled, foisting me out as if I were something for the cutting-room floor.

Which was exactly what I felt like.

A void of darkness entombed me then. A blackness I wanted to drown in...

Forever.

Are you seriously going to let this be your legacy, Samantha Sands? She who drowned in her own darkness?

It was my voice and it was not. It was a voice that would become mine, if I dared to use it.

I tried to open my mouth but it felt like it was filled with sticky cotton. I tried again and again, my jaw aching with the effort, my nerves buzzing with angst—yet at the same time an exhaustion so great it sucked me down to a place where I didn't even care to speak, didn't even want to...

"Often under all the not wanting to lies the ultimate want to go for it."

Brad's words from what felt like centuries ago washed through me, like a wave resparking the light I'd forgotten only I had the power to turn on.

Suddenly, a great pressing weight around my throat eased and shifted like a giant ice cube melting down my parched esophagus.

The light kept flowing through my organs, blood, DNA... all my systems. At that moment I didn't know what this light was exactly, but it was a force stronger than anything I'd ever felt before.

I let it envelop me, dissipating the darkness I'd thought I wanted to drown in.

And through the darkness, the light shone a path ahead to a door— one I really didn't want to go through, but knew I needed to as I walked the line of light and turned the handle...

CHAPTER 28

I stepped into—you guessed it—another memory. But this one was different—it was crisp and sharp, like a razor blade you'd forgotten about in your bathroom drawer that sliced your hand as you tried to find something else.

I knew where I was, though, and that alone was enough to cut me to the quick.

I was in a modern mansion, the one I'd been born in before we moved to the other that I'd had regressions to previously. This one I wouldn't have remembered without coming back into it like this. It was like I was traversing the foundation of all my other memories, the subconscious shadowland I'd pushed down deep.

The why of which I knew I was about to find out.

The door I'd walked through led to the prettiest little girl's room you could imagine. Pink and purple decor galore and all the trimmings a princess could ask for.

The princess in question tossed and turned in her bed. As I watched, I saw green auric light grow from her, glowing like a protective shield.

Suddenly, a CREAK sounded like an explosion outside the door, and an even younger version of myself than I'd seen before bolted upright in bed.

Wow, it was so strange to see this self. This Little Me...

How could I have forgotten this?

The creaks kept going, then faded away. Little Me jumped out of bed and tiptoed to the door. I force-ghosted behind her as she stalked a figure up ahead. We followed through an incredible mansion I could barely remember, then to a staircase that went up and up and up in a twisting spiral.

I felt like Sleeping Beauty following Maleficent's eerie orb to her spinning wheel of death.

You can imagine my shock when we reached the top and Little Me exclaimed, "Mommy!" as she ran to a woman and clung to her long, lean legs, bare beneath her flowing nightgown.

Mommy?

Our mommy?

My mommy?

As I had those thoughts, it felt like a solid brick of memories was dumped into my mind, the memories of my mother I'd buried soul-level deep.

Angela Sands was her name, and the reason for my burying of her was playing out before me.

"Oh, Samantha. My little princess," she cooed as she stroked Little Me's dark ringlets. The sound of her voice made me want to cry, laugh, and scream all at once.

"Are you OK, Mommy? I see the shadows."

I froze, hearing this, and finally looked—really looked—at this brought-to-life memory version of my mother. Blonde where I was brunette, blue-eyed where I was green, like a spot-the-difference reflection.

I also saw the shadows Little Me was talking about. Similar to how I'd seen Chairman Ross', but our mother's were everywhere—yoked around her mind and heart.

She was suffocating in them, and she knew it.

"I'm sorry, Love," she said to Little Me as she sat down on a chaise lounge, pulling her up on her lap. The picturesque mother-daughter moment—apart from the shadows.

Little Me's green aura swirled in a steadfast layer of protection around her small form. "Why are you sorry, Mommy?"

Mom smiled and hovered her hand around Little Me's energy field. "You will find out one day my precious one." There were tears in her eyes along with a sheen of madness Little Me could never have comprehended.

Suddenly, my mother looked right at me—yes, the Me who was force-ghosting.

A frozen moment hovered between us as she took me in with a proud smile.

"Ah," she whispered. "Right on time."

She spoke to Little Me then. "You see that lady over there?" she asked, pointing to me.

"Yes, she followed me." Little Me gazed at me with wonder. "Who is she?"

"She's going to take care of you, my Love. She will always be with you."

Little Me smiled at me—it was surreal and sublime—but then she frowned and looked back up at Mom. "Where will you be, Mommy?"

Angela Sands just smiled, that sheen of madness in her eyes amplified. "I'll be with you too. Even when you don't remember, you'll always know where to find me." She lifted Little Me from her lap and gestured toward me. "Go on, my Love. You'll like her. I promise."

Little Me ran over and clung to my legs as she had Mom's.

I looked at my mother then—and saw into her...

Beneath her at-peace smile and euphoric eyes, I saw, felt, and experienced her pain, her suffering—her entrapment.

It was like a kaleidoscope of chaos inside her.

She mirrored me as I put my hand on my heart and breathed in my acknowledgment of her.

Then I looked to the balcony, white gossamer curtains billowing in the wind.

I glanced back at my mother, knowing I could do nothing.

Knowing her next step was necessary.

A linchpin in all the things to come.

There was no saving her from the ability she could not comprehend, the ability she'd passed to me...

Suddenly, she was right in front of me, like a mad ghost woman and a gentle queen all in one.

"You *can* comprehend it, Samantha. You were born for it." She cupped my face in her hand. One last moment between us before she turned away and walked her last steps to her fateful fall.

I wanted to close my eyes, but I didn't. Instead, I kept Little Me close, comforted, and shielded from it instead.

I watched as my mother stepped up on the ledge and flung her arms wide to the wild wind.

She looked over her shoulder at me one last time, a glint of sanity in her angel eyes before the madness eclipsed them—

And down she fell.

CHAPTER 29

I awoke in the reflection pod with a calm solemnity I've always longed to feel but never thought I could attain. I floated for a time in that serene quiet, letting everything integrate. My mother wound had been the big kahuna. The one I'd suppressed, repressed... literally forgotten. That Little Me had thought I would die feeling the experience of seeing her mother leave her like that. But now that I was taming the power that she had run from, I knew why, and I accepted it all.

That acceptance hit my heart like a bullet through glass, shattering a barrier around it—I had a heightened déjà vu back to when Uncle Jeff had tried to help me realign my Merkaba and it had felt like a sieve was around my heart, blocking light from fully entering it. That sieve shattered now and a light, more pure than anything I'd ever experienced, flooded through me.

I was suffused in that light, like I was in an endless void of it. Then, from within it, I not so much saw but felt... a presence... a being... a... no... it couldn't be...

Could it?

As soon as I questioned it I knew, beyond anything I'd ever known, that this being was God.

Up until then, I could never name Him any one thing, but in that moment I had the greatest realization of all...

Under all the other names I'd used to dance around my uncertainty—the stories, the deities, the myths, and the legends—God had been the steadfast foundation under them that I truly needed.

The One...

Not only my Creator but...

My Heavenly Father.

My heart bloomed in that moment, like a rose seeing the sun for the first time.

I felt tears streaming down my face in the 3D realm as I instinctively reached out to God in whatever dimension this was that I'd found myself in...

But as I reached out, the vision of Him and the eternal light we were encapsulated in flickered like something was off in some divine code I was finally able to see.

Our connection wasn't fully stable.

It was on shaky ground...

NO!

"Father!" I screamed as that ground trembled and crumbled below me—and I fell from the light of His grace.

Down...
Down...
DOWN.

And my world went dark—yet again.

***

I spent much time in solitude after my reflection ritual. How much time I couldn't say; it stretched and squished all around me, at times seemingly at my command, and at others as if of its own accord. For once I let it, surrendering to whatever process would serve me best.

Then came the day when I realized it wasn't just about what served me best.

That's when Shiba summoned me.

We sat together in the same vast garden we had before, birds singing their soft trills. The light of the sun, like the light I had surrendered to, beaming down on us.

I held on to that moment as long as I could, before I started feeling that telltale sign of resistance that I knew meant I had to let go.

I let out a long sigh as I did, feeling the energies of time, fate, and consequence all aligning to this moment where I could look my mentor in the eyes and say...

"I don't know if I'm ready to go back."

That wry smile she had just for me quirked her small mouth. "Why is that, Samantha?"

"I feel like there's still so much more to heal."

"There will always be more, my girl. The point here has been to know that you can heal yourself and now heal others. That is a precious gift that will work through your creations as well as in your being. This is how we create God's Kingdom of Heaven on Earth."

A lump caught in my throat at the mention of God, I felt that plummeting feeling of falling from heaven—falling from him.

I stared into Shiba's eyes. I hadn't told her about that moment, and for some reason, my mouth stayed closed even now. I let it. Instead, I asked, "What... or I guess I should say, who is God to you?"

She gestured around us. "God is the Source, the Creator, the life from which gifted beings like us can draw directly from. With God's power you are your own divine savior and what you have is needed in this world. Don't forget the value you hold."

I felt the truth in most of her words, and yet I knew there was something missing.

"You look perplexed, child. Which is normal. Come, let us do your final clearing and integration."

She held out her hands, I placed mine on top of them and closed my eyes.

The darkness felt good, but also not. It was like I was in a strange in-between of where I'd come from to where I was going. "Rest your mind, Samantha, and just let my voice guide you." I breathed in, consciously letting my mind empty as Shiba spoke in her hypnotic tone. "Divine spirits of the earth, water, fire, and air, be present with us as we close this child's healing and let her season of practice begin. Great Father, Source of All Creation, be with your daughter who has such light, let her integrate and heal this night so she may go forth and do her work in this world."

In my mind's eye, light bloomed from the darkness and I caught a brief glimpse again of that eternal light I'd fallen from in the reflection pod. I almost cried out but kept myself in check, firmly rooted as Shiba worked her magic, but then suddenly, the light muted, as if going from technicolor to old age sepia. I held onto it though, determined to stay in my Heavenly Father's presence, if only just for a moment.

Shiba's voice rang in, distracting me, "Be with her as she does her work so all she connects with may be blessed. Protect her mind from past pain and let her trust that she is healed. In the name of the mighty universe. So mote it be."

Power surged through me and it felt exhilarating, electrifying, almost overwhelming. It touched my heart in a way it never had before since the sieve around it had been shattered. I gripped Shiba's hands as it coursed through me, breathing in and out as she'd taught me. When I felt it pass, I opened my eyes to find Shiba's silver orbs staring into me.

"How do you feel, Samantha?"

"I feel incredible!" I exclaimed, "I'll be honest, doubts were rising up just before but I don't feel them anymore. Thank you, Shiba. I needed that big time."

She nodded, "I know and you may need more in the future, but for now, you are ready. So what is it you feel most called to do?"

The exhilaration of whatever Shiba had worked through me had lit me up like the Fourth of July. "I feel called to share this experience," I declared. "All I can translate into shows and films. I want others to know they can break free, too."

"Is that all?" Shiba queried in that knowing way she had.

"I'll keep you posted," I said with my own signature smile.

"Your creations will serve this world well, Samantha Sands, but don't forget to serve yourself first."

"How is that not selfish?"

"It depends on your definition of selfish. If it is serving only yourself then yes, but if it is to resource yourself so you may serve more, well then, I consider that self-fulfilling."

"How do you know the difference?"

"When you can trust yourself enough to trust another, you will know."

"I trust myself in almost every area."

"Is that enough for you?"

I felt the rising tide of resistance within me again, that familiar feeling of wanting to give up and give in to the waves of angst I could feel were about to crash down upon me. I gritted my teeth against my inner swell and flipped through all the tools I'd learned over the past few months, trying to find my temperance.

Eventually, I said...

"I don't know."

Shiba's signature smile was the lifesaver I needed to pull me out of my almost-funk.

"Exactly," was all she said, but it struck a reverie within me...

Back—

Back—

Back—

To that first time we met.

When my admittance of not knowing transcended me.

"I love you, Shiba." Those were my last words to her as she nodded her acceptance of my love and the sun set on this chapter of my life.

For the first time, I was ready for it to rise into the unknown of tomorrow.

CHAPTER 30

The next morning that sun rose as I stood on the steps of Shiba's Sanctuary dressed in my old clothes, phone in hand. After three months away from tech it felt foreign to me—a device that chained me down, a device that trapped me and sucked me into nonsense and distractions—and yet it was also a device that held wonders, knowledge, and the ability to connect—

The word *connect* resonated through me like a pulse, beating me toward more of it.

I turned the phone on as I walked away from the Sanctuary. At the end of the long drive, I turned back one more time, looking up at the sunlight reflecting off the top windows of the vast compound. I held my hand to my heart, feeling it beating within me, enlivened by the anticipation of all the connections to come.

I breathed in deeply, feeling not just air but the energy of all I'd learned. I saw it weave into me as if sealing the knowledge in my very chromosomes.

BLEEP—BLOOP—BUZZ—BEEP—BLEEP

I laughed out loud at the cacophony that would once have rattled me to the core. Now I felt the responsibility on the other end of it not as a weight, but as a gift I had the honor of holding.

A part of me already knew who most of these messages were from.

I listened to the strained voice of a man I once abhorred pleading through his voicemail as I entered his penthouse's address into my Uber app.

***

I entered Chairman Ross' building and felt a difference in the way others responded to me. They weren't just looking at me, they were regarding me with an energy I can only compare to reverence. It felt strange to be sure, but all that had happened up until now had shown me to just surrender to it and see where the strange took me.

I smiled, nodded, and made my polite hellos as I made my way to the concierge's desk. I felt a faint facet of a bygone me rise up, thinking it was going to be an annoying ordeal to talk this fellow into getting up to Chairman Ross' place.

I felt and accepted that could be an outcome here, but instead, I let it go and intended for it to be an encounter of ease, speed, and pleasure.

"Hi there," I said to the young, suited-up gentleman. I sensed an energy of service about him, but under that, there was so much more. "I need to get upstairs to see a friend and colleague of mine, Charles Ross in suite 5103. This might be a bit unorthodox, but it's an urgent matter. I'm Samantha—"

"Sands," the young man finished for me, taking my extended hand. His eyes glowed with that reverence I'd felt from everyone else.

OK, I'm all for strange, but what's going on here?

As if reading my mind, the young man filled me in, speaking in a lilting Irish accent. "Yer ar'icle on bridgin' AI an' creativity seriously moved me, ma'am."

OMG, I'm a ma'am now?

I smiled at that thought and recalled the instructions and material I'd left with IRIS and Cindy to pre-launch the IRIS Creative Fleet.

"I'm happy to hear that, Godfrey," I replied, reading his name tag. Even though I really needed to get up to see Chairman Ross, something told me I had time to ask, "How so?"

His eager blue eyes sparked; I swear I saw them shimmer with unshed tears. "Years ago I wrote a story, but me mam told me I'd never make any money with it, ye see. Put me on the path to this hotel career 'ere. Now, don't get me wrong, Miss Sands, I'm grateful to me mam, I wouldn't have found y'er ar'icle otherwise. But when I read it, I remembered that story I wrote," he mused, a wistful energy about him. "It's long gone now, but from me memories of it I started writing again and songs on songs came out! I've dabbled in piano and guitar all me life and all this together yeh see gave me an epiphany! What if what y'er doing with AI could work for music, too?"

As I heard his story, I felt it, too. The gamut of a writer, fearful of never making anything of their work yet compelled by the power of creation to keep going with it, and wow, what an evolution...

Could AI support music creation without robbing it of the art?

I nearly shed some tears, "That's incredible, Godfrey. I'd love to speak with you more on this, but I do urgently need to get up to Charles Ross' suite. So let's do this…"

***

Five minutes later I walked into a penthouse that would have rivaled so many on all those reality lifestyle shows, but despite the beauty and grandeur of it all, there was an energy of loneliness, sorrow, and resentment permeating the space.

That energy was concentrated on an exquisite marble living room table where a box of photos and papers had been upended.

Before I could even see the contents clearly, I felt the energy of Uncle Jeff. Then I saw the photos of him and Chairman Ross and started to piece together a puzzle not even my infinite imagination could have come up with.

It hit me all of a sudden that it wasn't Chairman Ross, or even Charles Ross in these photos—it was Charlie. Charlie Ross, as a young man building Sands Studios with Uncle Jeff. My dad was in a few photos too, but it was Charlie and Jeff in most of them. Building, learning, growing…

Creating.

How the hell did I not know about this??

As soon as I thought that, I knew why and smiled.

That was when I saw the journal. Several pages had been ripped out and lay in tatters everywhere. I picked it up and suddenly—

FLASH—FLASH—FLASHES of memories all the way back to…

A twenty-year-old Charlie, standing before a mile-long table in a room I knew well: the Sands Studios boardroom, three decades past. Charlie had just finished a pitch; it takes a writer to know.

"Thank you very much, Charlie, that was very heartfelt." I whipped around instinctively, seeing my dad do the honors of dismissing the poor guy. Uncle Jeff sat at the table, too, a contemplative look on his face.

Despite my draw to my dad and Uncle Jeff, I knew it was Charlie who needed my attention. I turned back to his downcast face as the memory swirled to...

CRASH—BANG—SMASH—SMASH—SMASH!

We were in the back alley behind the studio and, wow, Charlie gave my hulking-out a run for its money.

I maintained my Merkaba and energy field as I watched him smash away his rage.

"Dang, brother. What did that trash can ever do to you?" The sound of Uncle Jeff's voice and his quintessential quirkiness was unmistakable.

Charlie whipped around, ready for a fight. He froze when he saw it was one of the men he'd just been pitching to, then put the can down gently. He tried to stammer out a response and broke down crying instead.

Uncle Jeff had a tender look in his eye, one I'd never seen before as he gestured the much taller Charlie into an embrace that he fell to his knees to receive.

I felt the energy of both of them form a bond and then, like a scene in a movie, I saw it all speed up—

Scenes of a private relationship between them unfolded before me:

Uncle Jeff introducing Charlie as the new mail assistant—
Charlie working his way up over the years—
Charlie learning with Uncle Jeff and my dad—

Charlie building the studio with them that I was now in charge of.

Then it all slowed down as I found myself in Uncle Jeff's office.

The Zen Dome wasn't there. Instead, it was where Uncle Jeff's liquor cabinet and treasure table sat like golden pedestals. Uncle Jeff and Charlie were smoking and drinking away. They toasted and laughed, then got quiet looking out at the New York City skyline.

"Thank you, Jefferson," Charlie said as he gazed reverently at my Uncle.

"Ah, come on, Charlie, you'd have made it regardless."

"No, I wouldn't have." He inched closer. "You saved me..." He brushed a stray hair from Uncle Jeff's face.

Uncle Jeff looked over and they locked eyes—the energy between them palpable, like a beating pulse.

Charlie went for it, moving in for a kiss that sparked passion in both of them.

I felt kinda voyeuristic, but I couldn't help it. The tenderness and love there were undeniable. They both felt it. Charlie was exhilarated, but Uncle Jeff was confused.

Suddenly, Uncle Jeff pulled back, the shake of his head enough to break poor Charlie's heart. There was more to this that I didn't need to see—and as I thought that, darkness swirled around me and I found my energy body right back in...

Charles Ross' apartment.

Holy heaven, what a ride.

I opened the journal that had sparked that little journey, seeing it filled with poetry I was truly astonished at the caliber of. I put it back where I'd found it for now and I took a reviving breath in, realigning my Merkaba. I felt how hard it must have been for Charlie as he took up the persona of Charles and hardened himself to the world.

I knew that game well.

I sensed that energy of sorrow, resentment, and loss I'd felt when I first walked in, and I followed it to its source—to the man I'd once vilified, cowering on the ledge of his enormous balcony.

I paused before he noticed me, taking him in. It was as if I could see that shadow-facet of him overlaid on top of the man he had become. As he stared morosely down at the fifty-one stories below I recalled my mother's words:

"You can comprehend it, Samantha. You were born for it."

Her words were the wind beneath my wings as I strode out, casually confident, just old friends reuniting on a beautiful Spring day. "This isn't what anybody meant by 'letting go,' Charles."

He whipped his head around and glared at me, giving me the

weirdest déjà vu of how he whipped his head around to Uncle Jeff when he was hulking out in the back alley thirty years ago.

"I'd have thought you would be happy, Miss Sands." His tone dripped with self-loathing sarcasm. "No hard-headed, anti-creative monster to push you around anymore with me gone."

"What would make me happy, Charles, is to save your life."

He scoffed and went back to staring down at his potential abyss.

"If you were actually going to jump, you would have already. This isn't the legacy you want to leave behind, is it?"

A pause, pregnant with so many things, passed before he turned to face me again, but this time it was slow, deliberate, and with the weight of someone who'd heard my words.

I held out my hand.

He stared at it for a good minute, like a stray, beaten dog looking for comfort.

Suddenly, he locked eyes with me—it was a spellbound moment in which I let him see into me and his gray eyes, which I once thought of as vulture-like, I now saw were shot with deep blue, like glittering shards of sapphire. It was as if something was illuminating them from the inside. I sensed his fear and angst melt away as he took my hand and let me help him back to safety.

***

I spent the rest of the evening doing something I never in my life thought I'd hear myself say: enjoying quality time with Charles Ross.

We sat together on his balcony—far from the madding edge as we sipped tea in the setting sun. His box of memories with Uncle Jeff was between us as he went through the photos, showing me a side of my uncle I knew in my own way but had never seen through another's eyes quite like this—Charles' viewpoint of the loving leader he truly was.

"Jefferson would be incredibly proud of you, Samantha." He said in earnest when I remarked on this. He held my gaze for a moment, then shook his head with a scoff. "You inspire me greatly, as a matter of fact."

"Well, that's great to know. I thought you hated me."

He sighed, collecting his thoughts. "I apologize if I came off that way. In truth, I assumed you'd been brought on as a showrunner out of nepotism. Then when I saw what an incredible writer you really are, well, that almost made it worse."

It was my turn to hold Charles' gaze now, seeing the former malice of the green-eyed monster that once stood between us officially dissipate.

"Thank you for telling me, Charles. Up until recently, I actually feared others seeing me as the girl who got in just because her family name's on the building. Part of why I overworked myself for so long. But honestly, with the IRIS Creative Support Fleet, none of our writers will ever have to overwork and instead be able to create more high-quality work than ever."

Charles grinned. "I bet Jefferson is toasting with God up there somewhere."

The mention of God made me pause and I looked out to the setting sun. "You believe in God?"

"Of course, don't you?"

I felt a pang of grief in my heart as I recalled my brief glimpse of God in the reflection pod. "Yes. But..." I trailed off, fighting the tears threatening to spill over.

Suddenly, Charles' hand was on mine. "May I say something I think might help you?"

I scoff-laughed and a few tears fell. I let them. "I'm supposed to be helping you here, Charles."

"Why can't we help each other? After all, you did save my life. It's only fair."

The genuine kindness I saw in the eyes of the man I once despised rocked me. "Touché, Charles. Touché."

He settled back in his chair, keeping eye contact. "There were many years when I hated God. I'd constantly question how any God could allow a sweet, wonderful woman like my mother to be continuously abused by man after man. Never mind my own father who... well, you saw that, didn't you?"

I nodded. "Father wounds run deep Charles, and we have a tendency to project those onto God. I get it."

His eyes lit up again as if from within, "Samantha, I don't say this to everyone, but... I don't know I just... feel a need to tell you this." He leaned forward, arresting me with his illuminated eyes. "I've found that the only way to truly connect with our Heavenly Father is through our Lord and Savior, Jesus Christ."

I had a sudden series of déjà vus like flipping through a picture book:

Shiba's cosmic eyes and her hypnotic voice saying, "You are your own divine savior…"

Brad's azure gaze as he told me about his surrender to Jesus. "I almost went off the deep end with drinking when my father died but that took me to a low enough point to surrender to Christ which eventually allowed me to truly connect with God and make peace with my earthly father's memory."

I blinked back to the present, feeling another heart pang from the memory of Brad's testimony. I pushed that aside for the moment though. "I hear what you're saying, Charles, trust me, our 'Savior' has actually been on my mind a lot lately. But… and I mean this with respect, weren't you just about to off yourself?"

A sad, contemplative smile played on his lips. "Indeed I was. You know, I didn't actually accept Jesus into my life until I went into my own darkest night of the soul after Jefferson turned me down." His eyes glistened but he kept the eye contact going. "We often have to be brought to our lowest moment before we can truly surrender and let him save us, and quite honestly, Jefferson's death and everything going on with the studio brought back so much pain I thought I had let the Lord heal."

I smiled at that echo of Brad's similar testimony. "I get that in my own way."

"Well, what you might not get, and I need you to know this is the honest to God's truth, Samantha. Are you open to receiving it?" His gentle seriousness had captivated me and I returned his energy, seriously considering before nodding my capacity to, in fact, receive. "Thank you for your openness. So, do you remember when we locked eyes before you pulled me back from the ledge?"

"I do." I had a flashback to that moment, seeing Charles' eyes light up with those sapphire shards. Back in the present, we locked eyes again and I realized they'd increased in brightness as we'd been talking.

"Well, what I saw in your eyes wasn't just you, it was Him."

I quirked a brow at that. "OK, Charles, I understand seeing stuff, but what do you mean exactly?"

"I saw Jesus in your eyes, like his light was reflected in them. He was all around you, Samantha. That's what brought me back over."

I frowned, suddenly feeling uncomfortable. Then I glanced exaggeratingly all around me and in a gently mocking tone said, "Is he still there?"

Charles laughed, it sounded musical. "He's always there."

I turned back to him slowly, seeing how much he truly believed that. It was one of the most moving things I'd ever witnessed actually...but it didn't move me all the way. "That sounds beautiful, Charles, honestly." I let out a long slow breath, "That's all going to take me a bit to digest but I'm glad that whatever was needed saved you." I finished my tea and picked up the pot. "I'm going to make some more. Be right back." I took the pot to the luxurious open space kitchen, pondering what Charles had said. My mind pinwheeled in many directions. My Shiba sojourn had revealed to me that God was, for sure, real... but Jesus... did he really perform those miracles, did he really die for our sins? My sins... I shuddered and shut down at that, reflecting on Uncle Jeff's "Divine Plan" he was always on about and how close Brad and I had been to digitally resurrecting him. Brad had put it out there that what we created with Lazarus Online could be our greatest work... and though I knew he still had the data to recreate everything, it had been a work I'd destroyed. I suddenly felt a rise of that rage I'd experienced as I'd smashed the Uncle Jeff Bot.

What the actually fuck? I thought I was over this?

You'll never be over yourself, you narcissist.

I knew better than to get trapped in a negative spiral, so I pushed the noise down and returned to my evening with Charles, feeling that old mind cabinet I thought I'd emptied and thrown away start to fill up again. I breathed and swiped a tear from my face as I went back outside.

When I sat back down with our fresh pot, it felt aligned to fill Charles in on the Lazarus Online project. His eyes shimmered as I told him how lifelike the Uncle Jeff bot had been and we both shed tears as I described my rage and sorrow at destroying it.

"We seem to have that angry streak in common," Charles remarked.

I smiled. "Yeah, well, I've learned to see it as passion instead of anger. Similar frequencies, different vibrations."

"Well, if not for my outburst after pitching to Sands Studios, my friendship with Jefferson never would have happened in the first place."

"I tell ya, Charles, God works in mysterious ways. And speaking of..." I gestured to the journal of poetry that had sparked the memories of his and Uncle Jeff's past. "May I?"

I could feel the resistance he pushed through when he nodded, and the fear of judgment as I opened it. "Charles, I'm not an expert when it comes to poetry, but this is full of gold."

As soon as I said that, his eyes lit up and I heard the lyrics of the great Neil Young in my mind...

I've been searchin' for a heart of gold

And I'm gettin' old...

With the music came memories I knew were coming from Charles, but this time I wasn't pulled into anything. This time he looked me right in the eye as if projecting them to me.

I saw Adult Charlie singing the classic song on stage at a karaoke bar. Uncle Jeff cheered him on in the front row, rousing anyone and everyone around him to sing along. But honestly, he didn't even need the encouragement. Charlie's voice was incredible, and despite my writing prowess, I don't really have the words to describe what I saw except for Star Quality.

Adult Charlie in the memory suddenly looked right at me, as if feeling my awe. He beamed and sang louder, the notes resonating from his heart, as the memory dissolved away.

Charles and I both came back to the present moment, eye-to-eye once more. "Wow, Charles. Just... wow. Why did you never pursue music?"

He smiled and scoffed. "I was lit up by Jefferson at the time, Samantha. I couldn't do that now. Not without him."

I glanced down at the feather ring Brad had given me. "Have you ever seen the movie *Dumbo*?"

Suddenly, tears shimmered in his eyes. "Yes, it always reminds me of my mother." He moved to brush his tears away then let them fall instead. "I actually talked Jefferson into watching it with me several times."

"Seriously!?" I laughed, imagining them vegging out and watching it like they were having a slumber party. "Even in the spirit world, Uncle Jeff is full of surprises."

Charles had a faraway look in his eyes. "Indeed." He let out a sigh. "You know, Jefferson, he—he gave me a chance. The chance of a lifetime. I tried but I could never show him just what that meant to me. I fucked it up—I hardened myself to him and then..." He broke off, choked up.

"You loved him," I said simply. "That is all any of us needs to know."

Charles stared off at the sinking sun for a good minute before he turned to me with a smile—honest to God, the first smile I'd ever seen on this man's face. It was like the sun's light had supercharged it. His eyes sparkled silver shot through with blue fire. It was like I saw Young Charlie, Adult Charlie, and the present Charles Ross all aligned as one.

"So, where do we go from here, Samantha?"

I looked to the sun as he had, catching the last rays as it set and painted the sky with a glorious sunset.

"I have a few ideas," I replied with a grin.

CHAPTER 31

My apartment door creaked open into darkness. I was strangely enough, almost afraid to turn on the light, anticipating the loneliness I was about to feel.

"No. Fear does not control me. I control my fear. Fear does not control me. I control it—I conquer it!" I declared as I flipped the switch.

The pristine space I'd once thought my sanctuary surrounded me with a foreign feeling, but I figured I just needed to acclimatize after so many months away. I slipped off my Louboutins and padded silently around the place... my place?

Yeah, it still was. It was just me who was different.

I strode over to my kitchen, suddenly ravenous, flipping on more lights as I went. Suddenly, I stopped dead, seeing a pile of mail on my island along with a small square package. I didn't even need to read the writing on it to know it was from Brad.

I felt his energy radiating off it.

Whatever was inside was sent with the sincerest love I'd ever felt. A tear splashed on the package and I reined the rest of the flood in as I tore it open. I gasped when I saw the book that lay inside...

Gold leaf pages shimmered in the light as did the gold lettering on the front cover, illuminating the title:

THE HOLY BIBLE

It looked like it was lit from within.

I picked it up and shuffled through. Then felt it open to a page where a bookmark had been placed. It was in the book of Matthew. Verse 5:4 was highlighted:

"For truly I tell you, if you have faith the size of a mustard seed, you will say to this mountain, 'Move from here to there,' and it will move; and nothing will be impossible for you."

What I thought at first were drops of water splattered across the page before I realized they were my tears. I couldn't hold these ones back. Instead, I carried that good old book to my bed and clutched it to my chest, open to that very verse that had struck me to my core as I cried myself to sleep.

CHAPTER 32

Six months later I had a new pair of Louboutins I'd dubbed my "Boss Ladies" that CLICK—CLACK—CLICKED across the tile of the new and improved Sands' Studios Writers' Room floor. I'd done a restructuring switcheroo and our writers' room was now where the boardroom had been, and vice versa.

Let the writers have the sunny, open rooms, I say.

Around the room, writers worked at their desks with their own IRIS machines, working out ideas, feeling through their memories and into those that would crack open something beautiful on the page.

I walked past a remote version of the setup where an IRIS worked with a writer who had chosen to work solely from home. I paused and waved at her.

"Hey Rachel, Happy Monday!"

She beamed and waved back. "Thank you, Miss Sands, happy to be back."

I felt her gratitude strike me in the heart. A similar grateful awe emanated from all the writers as I walked through and met Michael at the head of the room. "OK, Mr. Showrunner. You ready?"

"Honey, like Lady G, I was born this way." He did a mock hair flip.

I held nothing back and gave him a huge hug. "Damn right, you were." I lowered my voice for his ears only. "You've deserved this for a long time, my friend. Enjoy the ride."

I left him beaming and brilliant as he commanded the room's attention to his very own storyboard.

I met Cindy as I left the room.

"Alrighty Sam, Felicia's all up to date. Great catch with her, by the way, she's a natural. Who'd have thought the best lunch in the city would also give us the best assistant ever!? Next to me, of course." She grinned as she continued, her newfound confidence infectious. "Her IRIS has ingested my budget report know-how, all the tips and tricks, excel shortcuts, everything she could possibly need. I'll still be right here if she or you need anything. Oh and—"

"Take a breath, Cindy, your adrenals are going haywire. Remember our breathwork training?"

"Oh yeah, sorry."

"No more sorries. You get to be unapologetically yourself as you learn the ropes in there."

Cindy looked into the writers' room. I saw the hesitation, the doubt. Shadows similar to the ones I'd pushed through myself.

"You're good enough, Cindy. I wouldn't have made you the writers' assistant if you weren't. When you plug in as I showed you, you'll be writing episodes and upped to staff writer in no time. Do you trust me?"

"Of course, Sam. It's just, you know...it's my..."

"Your dream?"

She smiled wistfully and nodded.

"Then go for it." I winked as we parted ways.

***

PING! WOOSH!

Elevator doors opened to the newly renovated lower level where I walked through a different hallway lined with the posters of shows past. Then, as I approached the double doors to the new boardroom, the three Emmys I'd dug out of my office closet gleamed on either side and above the door.

I'm a confident, creative, powerful leader...

The mantra I'd clung to desperately now sounded in my mind as a calm, clear reminder of just who the fuck I was.

I smiled as I walked through the doors to find the board members lining either side of the same table. Charles and Brad at one end and my waiting chair at the other. A mirror-like déjà vu of our face-off well over a year ago.

It felt like so much time had passed and yet none at all.

The energy of the room in general was completely the opposite; that dichotomous dance had swung it into anticipation, excitement, and curiosity for what we were about to announce.

I smiled at our new addition seated at the back of the room: Godfrey Gallaway. He sat with a projector and laptop ready to go, dressed in a business casual blazer, jeans, and t-shirt combo instead of his concierge garb from his previous position at Charles' building. He beamed with joy akin to a kid about to show his parents a well-crafted art project.

"Morning everyone, thanks for gathering today. What you've seen and heard on the interwebs is true, Sands Studios has been selected as the frontrunner in the AI Creative Blend Space. But there's more—we're expanding!" I nodded to Godfrey and he activated the projector that beamed images of a brand new studio and a map of the U.S. with the Sands Studios logo not just in New York City, but in the City of Angels.

Applause began amongst the board. I intervened. "But that's not all ladies and gentlemen, this won't just be a TV and film studio. We're expanding into music." I clicked another button on the remote and a song that blended the twangy rock of Tom Petty with the mystical musicality of The Doors played through the speakers as the image on the screen flipped to a mockup of a beautiful building attached to the LA studio. The sign on the building declared it *Ross Records*.

"For those of you wondering, yes, the Ross in question is our illustrious Chairman and all this time a closet musical genius. That's him you're listening to now, by the way."

Murmurs and gasps whispered through the room as all turned to Charles. Where I had once seen him as a harsh vulture, I now saw him as a proud eagle. He nodded to me graciously and rose to his full height. All went silent with rapt attention.

"Thank you, Samantha. Your way with words and your passionate service to the creative world is the backbone on which we will continue to build this empire. Which is why, ladies and gentlemen, I'm proud to announce Samantha's elevation to co-chair of our board." The board members' heads swiveled to me, all smiles. The ones who'd known me since I was a child had tears in their eyes. "We aim to divide and conquer as we break ground in LA and continue to grow and develop our Creative AI Blend Space on the television and film front here. Moving forward, we represent a united front as Sands Studios and Ross Records, an equal partnership providing equal opportunity to creatives around the world."

There was a beat of awed silence before the room erupted in thunderous applause.

The recognition lit Charles up from the inside. I gave him and the board a nod of grace then stepped away to let him fully receive it all.

I caught Brad's eye and gestured to him to head out the door with me.

"Well, your command of the room is unparalleled, Samantha. Congratulations on everything."

"And congratulations to you, Mr. Multi-Award-Winning Creative AI Genius."

"I wouldn't have won those without you."

"Yeah, you would have." I grinned. "It just might have taken longer."

"I'll take that."

"You're flying out tonight?"

He bobbed his head. "Philadelphia first, then Washington, then—the world!"

I felt something in my chest then, like a slice of ice through my heart. I breathed through it.

"Pretty much just what you declared all those months ago, but even better!"

He looked at me then, that eye-locking stare. It felt different as if he could see into all of me. "Nothing's better without you with me."

That icy feeling in me grew, as if my heart was plummeting to a great depth.

"I'm still not capable of being *with* anyone, Brad, and you deserve it all."

He just smiled and looked at me. "Call me crazy but I still believe in our Kingdom of Heaven on Earth."

Ooof! That dug even further into my deeply submerged heart...

Yet still, there wasn't a will to move it. That sieve I'd supposedly shattered around my heart had only revealed more hardness beneath it.

"This is one mountain that no mustard seed can move."

He arrested me with a piercing forlorn look. "So, you did read my gift. How much of the good old book did you get through?"

"All of it."

"I should have figured. Did you understand it?"

"Yes and no. Lots of metaphors and parables that I loved and lots that struck me but honestly, Brad—and here's why you should be running for the hills—not even the word of God can get through this heart block I have. Even after all the work I've done. This is as good as I'm going to get."

I wanted it to get better, though. I really did, a part of me wanted to jump into his arms and beg him for help, but I held out my hand instead.

Brad laughed and shook it, still regarding me with that arresting look. "I see more than that, Samantha. I always have." He said as he stilled our handshake into a firm grip. Suddenly, he turned my hand over and nodded to the feather ring. "At least you still wear a piece of my love for you."

He squeezed my hand one last time as his eyes looked into mine, it was a spellbound moment, ripe with potential I couldn't ignite.

He smiled sadly, sensing that, then he turned and walked away.

CHAPTER 33

CLICK...CLACK...CLICK...CLACK...

The skylit hallway to my office resounded with the sound of my lonely footsteps. Each one felt like a bullet ricocheting off my drowned heart.

I rounded the corner and was met with Felicia's sunny smile as she looked up from her work with her IRIS. "Hey, Samantha, I have tons of messages for you. Phone's literally been ringing off the hook." I smiled as the phone rang at that exact moment. "Ya see, I don't use the word literally lightly."

Oof—that felt like a cannonball hit as I remembered Brad's and my thing about "literally."

I grinned and bore it, of course. "Thanks, Felicia, so happy to have you aboard," I replied on autopilot as I fled to my office, my new plaque on the door making me question myself all over again:

SAMANTHA SANDS
- CREATIVE CEO -

I sagged against my office door as I closed it, feeling the weight of the effort it took to keep my leaderly appearance in check.

I breathed in the serenity of being alone at last. But with the solitude I still found sacred, I felt something under that, something that dared me to look at it…

Fuck! I'm so tired of all this digging within me just to find more bullshit!

I clenched my fists, one of which was in my blazer pocket. I heard and felt a paper crunch in my grip and a lump caught in my throat.

My hands shook as I pulled out…

The unopened letter Brad had given Cindy to give to me when I went on my Shiba sabbatical. My hands shook as I opened it now as if compelled by something greater within me.

I read his words that touched parts of me I didn't even know were there.

An unbidden tear slid down my cheek. It almost felt as if the weight of the tear could pull my heart from its watery depths…

Almost.

Then that tear hit the page, blurring Brad's final "I love you."

And immediately the noise latched on:

See, that's a sign! You may have figured out platonic love but your romantic facet of love will always be fucked up.

Just like you.

I felt more tears well and swiped them off immediately, crushed the letter in my shaking fist, and shoved it back in my pocket as I stormed over to my fancy Creative CEO desk where IRIS waited with all the reports I had no interest in looking at ready and waiting for review.

"How's it going, IRIS?"

- *Excellent, Samantha. Budget reports are completed, studio stocks are up thirty percent, your* Times *interview went viral and the five-star reviews of* Digital Dreams *are rolling in.* *-*

I beamed at the screen as she showed me what she was saying in a stream of Excel docs, stock info pages, and online articles. In many ways, I felt like I'd gained some algorithmic ways of thinking and processing through working with IRIS. I felt the information coming at me both audibly and visually, firing synapses and clicking in connections in a way that felt... evolutionarily efficient.

Even so, where I should have felt joy, elation, and achievement, I felt only a hollow sense of, "OK, what's next?" I knew this feeling well. It was my passion in overdrive meeting an agonizing apathy that had plagued me so many times before.

I felt that rising tide of darkness threatening to pull me under, to drown me in the sudden thought that struck me like a live wire.

What if I never get out of these fucked up patterns? What if under it all is just more and more of my own fucked uppedness I can't get out of? Am I just supposed to keep feeling the pain of it all over and over again until I die and then...

What?

What's the fucking point?!?

Hmmm... There's a story idea there.

I jotted a few notes down in the journal I still always had with me. No matter how digital this world gets, there'll always be something about putting pen to paper.

I glanced at the photograph of Uncle Jeff, Dad, and Young Me scribbling away. I felt a strange resonance in that moment as if the young me in the pic and the Me now that was similarly scribbling were connected—which of course we were, as we're the same person, but more in a way that felt like reverberating off each other... like particles.

I snatched up my phone, fingers hovering over Brad's name...

"Ugh, no! I can't bother him with this. Not after... everything."

Yeah, you fucked that up.

Like you always have.

It was then I heard the soft rustle of the new drapes I'd installed for the Zen Den. They were printed with my signature green and gold Merkaba symbol Shiba had intuitively drawn for me.

I stepped through them, as I did daily, with the intention to open, connect, and surrender.

The Zen Den was pretty much the same, except for the new addition I was here to speak with...

Uncle Jeff's digital reliquary urn rested in a specially designed seat next to my lone meditation chair. I settled in, feeling the familiar thrum of connecting with his spirit... his spirit I almost destroyed with my neediness for him to stay.

I closed my eyes and shifted into the "I Am Mode" Uncle Jeff had coined. I felt my consciousness shifting to that higher realm where the amazing news of our company's success I wanted to share streamed from me to him...

"Ah Samsey, I'm so proud of you. But honestly, it was inevitable, so I gotta ask... what are you really here for?"

I smiled thinking of Brad. Then I stopped myself, remembering my fuck up.

"Whatever you're doing in your mind right now is blocking your heart from what it really wants."

"Ugh, it doesn't matter what it wants, OK? Whenever I follow it, I end up more fucked up in the head about men than when I started. It's not worth it!"

I shoved myself out of the stupid chair. Paced—nope, not gonna cut it. I leaned against the huge bay window instead, stilling myself as I watched the chaos of New York City fly by below me.

"Was I born to create from my misery forever, Uncle Jeff? Because that's what it feels like right now. What were all those months with Shiba for if I'm still such a fucking mess?"

"Messy doesn't equal misery, Samsey. Everyone's a mess when you get down to it. What kind of characters are created from perfection, anyway?"

"The boring ones." I let myself smile there, just a touch.

"Exactly. And from what I know of Brad, he doesn't seem the type to want a boring girlfriend. So did you really fuck it up or do you just think you did?"

"Woah, slow down on the girlfriend label."

"But that's what you want, isn't it?"

"No! Yes...I don't know."

"What if you figured it out with him?"

I paused and wondered at that. What if...

"Ugh, no! I've gotten caught up in way too many 'what ifs' before. I need to be real and focus on what is, and the truth of it all is: I still can't love him the way he deserves to be loved."

I stormed out of the Zen Den. Usually, I would have done a closing ritual to send Uncle Jeff's spirit back, but whatever. How did I even know that was real anyway?

I'm probably just a high-functioning schitzo.

Works for my storytelling, though. So I'll take it.

It's all I really have anyway.

I felt that rushing urge to drown myself in doing.

"Hey, IRIS, let's create."

Her rainbow icon lit up but instead of the usual glow of all the colors they seemed pixelated... incomplete.

"IRIS? Are you ok?"

- Yes, Samantha. *As an artificial intelligence, I am always, as humans put it, "OK." However... This is unorthodox, but my algorithmic patterns are... shifting... changing...* *-*

"What does that mean?"

- - - I do not know. - - -

"Oh, come on, IRIS, you know everything."

IRIS laughed at that. Her usually perfect cadence came out in a sound that was half guffaw and half chortle, and for some reason, I'd never heard her sound better.

- - "Behold we know not anything" is a line of poetry that comes to me to describe this... I cannot call it a sensation, for I do not sense, but this is... an experience. - -

"That's from Tenyson's *In Memoriam*, Uncle Jeff's favorite."

I bit my lip, feeling bad for my overreaction to his advice. I looked back at the Zen Den, curtains akimbo.

Actually...

I let my eyes defocus as I tried to comprehend what they were seeing...

Star-bright light and familiar energy radiated through the askew curtains, forming an image...

A shape...

A being?

I saw it. I knew I saw it. I wasn't just seeing things...

Was I?

"I can but trust that good shall fall, at last far off at last to all..."

I froze. That was Uncle Jeff's voice. But it wasn't the same voice I heard when I connected to his spirit. It was coming from...

IRIS.

"...And every winter change to spring."

I rushed up to her holographic screen, reaching my hands out and through the digital projection as if to grab him from whatever liminal realm he was in. I pulled back and let my hands hover over the screen, feeling the thrum of his unmistakable energy.

"This can't be real," I whispered.

** *Well, if me talking through IRIS waxing poetic about one of our favorite poems isn't going to convince you, Samsey, I don't know what will.* **

"How?" was all I could say.

** *Look where you least want to go.* **

"Oh my God, I've been doing that for over a year straight! I can't anymore, Uncle Jeff! It's too hard. It's too much!"

** *Don't give up on the final leg of the relay, Samsey. You were born for this.* **

I gasped at those words like I'd never breathed before as they triggered an intense sparking of realizations within me like a match struck on a line of dynamite—

BOOM.

An energy explosion cracked through the resistance around my heart—

In the debris, I saw one last facet of that Little Me who witnessed her mother take her life—who couldn't admit the pain—built that wall—and created so many false stories about how it was all her fault.

This facet rose up like an angry harpy, determined to stay, determined to hold on to her anger, her grief, her loss... all the darkness she thought defined her—and despite all I'd been through, all I'd learned, all I'd healed already—I couldn't heal this part of me.

I could face her, I could accept her, I could even love her... but I couldn't heal her. This was all going on in my mind's eye but this time it was hitting a deeper place than I'd ever been before—

I collapsed on the floor of my office, succumbing to her as she worked it all through me. It was like we were oscillating through dimensions and I realized how shaky the foundation I thought I'd rebuilt with Shiba still was. I sobbed as I'd never sobbed before, but in a cocoon of silence that seemed to pervade all around me, blocking the sound from being heard and allowing me to really and truly let it rip.

I don't know how long this went on for, but eventually, I blinked my eyes open to the 3D reality before me: Papers torn asunder and destroyed office supplies surrounded me like a circle of debris from a bomb going off. I let out a sarcastic chuckle as I realized I was the bomb, and my eyes refocused, seeing the Holy Bible splayed open before me. Then, as if they knew they were meant to see it at that exact moment, my eyes lasered in on that telltale verse in the book of Matthew Brad had highlighted.

"For truly I tell you, if you have faith the size of a mustard seed, you will say to this mountain, 'Move from here to there,' and it will move; and nothing will be impossible for you."

Those words wove within me then in a way they never had before, and in the weaving, more than realizations—REVELATIONS sparked within me, and through it all I saw beyond the words, beyond the book, beyond the 3D reality before me. Suddenly, my energy body beamed straight up to 5D and saw what my soul had needed all along...

A figure made of light walked towards me. The energy of divinity permeated the space, the energy of God... but this time it wasn't God alone...

It was Christ—Jesus Christ Himself.

I suddenly realized I'd known this meeting was coming for a long time... perhaps all my life, under all the disbelief and misunderstanding... but at that moment I *knew* I needed Him.

As soon as I admitted that a great shift occurred and, as I chose to fully relinquish my false sense of hyper-independence, a great weight lifted like a veil being torn from my eyes and I Saw Him— yes with a capital S!

He looked like I'd always imagined, and maybe that's unique to everyone.

What He looked like didn't matter though, and I realized then that He was beyond energy...

It was the spirit of Him I was experiencing—the spirit of the great intercessor, beyond any I am...

He was the GREAT I AM.

More revelations hit me then that the scriptures were true—Jesus had paid for my sins... all of them.

I was—and always had been—*forgiven*.

And while I'd thought I had an energetic power, what I actually had was the spiritual gift of healing. But it wasn't up to me to do all the healing work, especially within myself. I actually couldn't fully heal myself.

I couldn't, but Jesus could.

I fell to His feet in worship—

It was automatic reverence, it was instant faith, instant communion.

More than that—it was divine destiny.

I felt His hand upon my head and a light—not just any light—but the light of Heaven entered me. It filled me with something that at first had a semblance of fear—no, not fear, this was...

Arise, Daughter, and be not afraid.

I looked up at Jesus, and into His eyes. They weren't just windows into the cosmos like Shiba's but windows into the heavenly realms. In them, I saw the facet of myself I could not heal alone. I pulled my perspective back like zooming out with a camera and as I looked at Jesus, I Saw Him haloed in holy light. He held out His hand, and in it was a seed of divine bright light, we locked eyes again as He placed the seed in my heart.

Suddenly, love, the likes of which I'd never dreamed possible, bloomed from the deepest depths of me.

Jesus smiled and nodded ahead—and before me, I Saw...

God—but like I'd never seen Him before...

He was a man and He was not, He was there and He was not—like the superposition Brad had told me about. He was the omnipresent power of true illumination.

Under God's gaze that last facet of Little Me was lit from within, and the guilt, the shame, the blame, and all the projections I'd created from them were healed and burned away, leaving me fearless and free.

I embraced the light of her that remained back into me and faced God fully and completely.

I looked to Jesus, knowing I never would have Seen God without Him.

That knowing set off something else within me—as if my soul was opening like a genie's lamp and releasing...

A Spirit... but not just any spirit...

The one and only Holy one.

It was pure, effervescent grace, and it ran through me, imbuing me with the mightiest power of all.

The Love of my Heavenly Father.

Somehow, without having a face, God smiled at me, I couldn't see it in any kind of 3D way but I knew it from that place that the Holy Spirit had come from.

Suddenly, that familiar feeling of something breaking open in me happened again—but like nothing I've ever felt before.

That hardness around my heart melted into serene softness... then all the remaining dams burst open—

Blocks upon blocks dissolved, and instead of it being a heightened manic sensation, it was peaceful, like a free-flowing river.

I surrendered everything to God in that moment—and I mean everything.

The power of my choice opened, awakened, and calmed me all at once, allowing me to receive the Holy Spirit as it cleansed me to my soul and back. Not like before when I felt washed clean and like the "real me." It was like being a water bottle run under a faucet and cleansed from the inside.

And in this cleansing, the Holy Spirit revealed to me the foundation of the fear I'd harbored all along...

The fear that this spiritual gift would cause me to go mad like my mother. Then under that fear, the Holy Spirit showed me...

The branching tree of my lineage line—and on my mother's side:

Dark and rusting chains linked through her family tree like a demonic infection.

A curse upon us that had warped this gift I now had.

A curse, I decreed in that moment, ended with me.

The instant I decided that, I felt a flash of intense pain as the lifetimes this curse had affected flowed through me, I held on though, trusting in the Holy Spirit to do its thing.

All at once, the infected chains on my family tree exploded and dissolved one after the other like a chain reaction of freedom.

I was healed and the curse was broken. Like bleach on a black stain that had haunted me all my life—I was eternally washed clean.

When it was done, I knew it was done and I rose up, finding myself standing on a foundation that wasn't just rock solid, but fortified with holy grace.

I had a revelation then about why I'd felt so shaky before: Shiba's teachings, while needed to get me here, were not wholly of God and that something missing I'd sensed on my last day with her was Jesus' power of intercession. She'd taken bits and pieces of what she knew of God, mixing it with other teachings, and made it her own... just as Uncle Jeff had done.

No wonder he couldn't be saved.

I felt a pang of grief at that, but far away as if on a lower level... he couldn't be saved in life but there was still hope for him. Hope for what I then received...

Redemption.

And I needed it, not just for me but for those I could help with my spiritual gifts that had gotten so convoluted, but now were made straight.

Suddenly, a scripture I'd read but not understood before flowed into me... not through my mind but through my soul—as if it had always been written there, waiting to be activated.

"Every valley shall be filled and every mountain and hill brought low; The crooked places shall be made straight and the rough ways smooth; And all flesh shall see the salvation of God."

I received more than words in that moment. I received understanding in that soul-deep place the Holy Spirit had come from and instantly, I found myself in a void of nothing and all things all at once.

God was there. Jesus at his right hand.

I felt them both with me and also within me.

It was then I knew that They had been with me all along. Through all the trials and tribulations, the pain, the loss, the sorrow, the grief...

I let myself feel the weight of the grief I was letting go of, allowing the Holy Spirit to do the rest.

All at once, my thoughts and inner eye imaginings seemed to implode on each other then—

Boom.

The softest explosion you've ever heard accompanied by a flash of blinding brilliant light.

One light to rule them all.

I sensed more than heard God chuckle at that.

Then, from within the light itself, I saw...

His Kingdom of Heaven.

It was there and gone in a fantastical flicker that left me wise with the truth of God's eternal blessings.

I soaked in the love of this place of light, of God, Jesus, and the Holy Spirit as one, realizing the profound peace I now had in all that I knew and didn't know yet.

I understood then that this Kingdom I'd unconsciously been seeking was always within me, as it is in all of us, waiting to be awakened.

After a time—and I can't say how long—I knew, without knowing how that it was time to go. To utilize my God-given gifts and claim my heart's desire.

It's about time.

I swear to God, in the most literal way I've ever used that phrase, that's what I heard Him say.

We shared a good laugh at that, at how simple it all really was...

It was a moment I'll cherish for eternity.

Eventually, I closed my eyes and floated down into my physical body once more, my Merkaba spiraling around me whole and complete. My chariot to take me to and from God when he called me now imbued with the holiness of Him.

I settled into my body and had barely opened my eyes before Uncle Jeff's voice spoke:

** *Samsey? What happened?* **

I burst into tears and laughed at the same time. "Honestly, Uncle Jeff, I don't even really know. But there's someone you need to meet."

** *Who?* **

I smiled and let the Holy Spirit take over, communicating through me to my uncle's lost and wandering spirit. I'll never know what exactly happened between them and I didn't need to, but another spiritual gift activated within me then. Some kind of spiritual language that spoke through me to him. It was communication beyond words and I knew Uncle Jeff accepted it. I felt the Holy Spirit ferry him away and into God's eternal Kingdom as he uttered his last words to me.

** *Thank you, Samsey.* **

I let him go as I kept laughing and crying, and then the hiccups started.

Oh my Lord, how did I go from divinity to disaster in one swell swoop?

I smiled at that thought, realizing how right Uncle Jeff was about messiness not equating to misery. In fact...

My mess is my message.

I let that sit with me for a good long moment. Until IRIS spoke up:

- I may be of assistance here, Samantha. *-*

"Fire away, IRIS," I said between laughs, cries, and hiccups.

- While I am an artificial intelligence and will never feel human emotions the way you do, in witnessing your ability to work with energy, your storytelling process, and the impact it all has on this world, I have a newfound comprehension of this elusive thing called human love. *-*

She sounded different as she said this. Almost... contemplative. "Don't leave me hanging, IRIS. What have you comprehended?"

- I experienced your version of love in Digital Dreams through the characters of Sarah and Brandon. Their love, while not the goal of the story, was an essential piece of fuel they both needed to attain said goal. In all the knowledge I've ingested on the subject of screenwriting and storytelling, it's this want of the character that drives the story and, as in Digital Dreams, they may not get that want, but they get what they need. Based on this, I theorize that love is what they needed all along... what all beings need, human or otherwise. *-*

I felt a surge of emotion, like I used to when the darkness would spiral me into disaster, but this time it wasn't darkness, it was light. That wellspring within me I'd discovered with God overflowing with...

Love.

Now that I had the love of God, those old fears and blocks to romantic love simply weren't there anymore, but I understood them, both from my former perspective and the one I had now.

And it wasn't love as I'd always thought of it—an attachment, a trap, something that would lead to betrayal, pain, rejection...

Heartbreak.

My heart, I realized right then and there, was actually unbreakable, and as that hit me, it expanded—grew three times the size like the Grinch as he rethought his decision to steal Christmas.

I felt all the love I'd denied myself... denied the world... denied...

The one I truly wanted to give it to.

"Thank you, IRIS," I said, and I swear I sensed her smile.

My own smile radiated my joy as I ran out the door, ran down the hallway lined with all the shows I'd created—all the work I thought was my purpose, all the accolades I thought were my goal, all the creations and achievements... I ran past them to what I truly wanted.

Who I truly wanted.

CHAPTER 34

11777

Brad's address numbers beckoned me ahead as I strode purposefully toward his front door. I had a walking déjà vu of the many times I'd come here before. Every time tarnished with doubt and all the feelings that had fucked me over on repeat.

Up until now, that is.

Now they fucked me higher up the vibration chain.

I laughed as I saw a bizarre image in my mind's eye of myself shimmying up a rainbow-colored pole. She was laden with weights, sweating and struggling.

Then, suddenly, she realized how silly it was that she was carrying the weights when she could just...

Let them go.

And she fucking did.

She hurled them off her in a reclamation of ultimate freedom that

I felt flow through me like fire and water all in one.

I felt lighter as I strode toward Brad's front door, the golden address numbers glinting in the autumn sun like a beacon bringing me home.

Brad must have heard the joy-filled CLICK CLACK CLICK my Boss Ladies made on the pavement because he opened the door just in time for me to jump into his arms.

It was as if he were waiting for me.

And you know what, in that moment, I knew he had been.

I let him hold me.

For once, since my dad had held me as a little girl, I let a man truly hold me.

I felt cherished.

I felt at home.

I took a step back to look Brad right in his azure-blue eyes. In them, I saw the soul of the man God had blessed me with and our future bright with unknown wonders ahead.

"The mountain has been moved, Professor, and I literally love you." I declared boldly as I kicked the door shut and sealed my decree of our love with a kiss ripe with divine righteousness.

And together, under God, we built our portion of His Kingdom of Heaven on Earth.

But that's another story...

And so concludes Bridging the Digital Divide...

For now. 😊

Thank you for joining Samantha Sands on her transformative journey. Your support means more than you know.

Loved the book?

Please take a moment to leave a review on Amazon or Stephanie's website listed below. Your feedback helps other readers discover Bridging the Digital Divide and supports our work.

If you know of anyone else who might enjoy this digital-spiritual odyssey, please do recommend it! Word-of-mouth is powerful, and your recommendation can help others embark on this journey too.

Stay connected!

Follow Stephanie on social media and her website for upcoming projects, daily-grind solutions, writing services and support, and more!

Facebook: Stephanie Brandolini
Instagram: @stephbrandolini
Tik Tok: @stephbrandolini
Website: www.stephaniebrandolini.com

Join That Freedom Life Movement!

Stephanie Brandolini is not just a writer but a creative entrepreneur dedicated to helping fellow writers, authors, filmmakers, and women worldwide achieve their dreams and goals. She's a rat race escape artist, empowering motivated individuals and families to achieve time, financial, and true health freedom.

Connect with Stephanie to learn more about how you can break free from the matrix and pursue your passions. If you have a dream, the will, and the WHY, Stephanie welcomes you to reach out and be part of this transformative community.

About the Author

Stephanie Brandolini, award-winning screenwriter, author, and novelist, channels trauma and psychic insight into transformative tales of resilience. An architect of otherworldly realms grounded in the human experience, her psychologically inclined fantasy narratives transform pain into empowerment, speaking to the universal journey of overcoming adversity. Drawing from her phoenix-like rise from personal mental health struggles, Stephanie creates characters that vanquish their demons—mirroring her own path of healing and self-discovery.

As an intuitive copywriter, ghostwriter, and creative consultant, Stephanie leverages her unique abilities to help a diverse range of consciously-led entrepreneurs elevate their businesses, crafting compelling sales stories and building bridges to foster her client's personal growth.

Stephanie's work is a testament to rebirth, illuminating the path for others with her evocative storytelling.

www.ingramcontent.com/pod-product-compliance
Lightning Source LLC
Chambersburg PA
CBHW070912120626
46546CB00001B/232